Acupuncture & Infertility

By Li Zheng and Sara Allen

Edited By Changhong Zhou

Published by Lulu, 860 Aviation Parkway, Suite 300, Morrisville, NC 27560
ISBN: 978-1-387-95260-1
Requests for permission to make copies of any part of this work should be mailed to:
Li Zheng and Changhong Zhou
105 Chestnut St., Unit 36
Needham, MA 02492
Website: www.bostonchineseacupuncture.org
www.allenacupuncture.com
E-mail: li.acumagic@gmail.com
sara@allenacupuncture.com

Dedicated to my parents and my three healthy children: Iris Zhou, Steven Zhou, and Darren Zhou

Acknowledgements

I would like to thank my patients who trust me and gave me the opportunity to use five thousand years of acupuncture and Chinese Medicine to help them create a healthy wonderful life. Each patient brings their own specific personal experiences and journey to fulfill their dreams in their life. I've learned a lot from their stories over the years and have been using scientific research to analyze success stories and pass that knowledge onto other patients, so they can save time, money and reduce stress when they are trying to conceive naturally or through IUI and IVF.

I would like to thank my daughter, who is a law student in Harvard Law School. She urged me to finish writing the infertility pamphlet and book so I can use my 26 years of clinical experience to help women maintain good health before, during and after pregnancy.

Foreword

Legend has it that a few thousand years ago, somebody bumped into a pointed object and found their ailment cured. Since then, every inch of the body has been explored to find the 'points' that treat various health issues. Amazingly, this treatment is all accomplished with thin needles.

Such is the case of acupuncture and herbs with fertility management. Whether due to physiological or psychological elements, there are many scientifically-constructed studies that show benefits, meaning the observed increase of pregnancy rates in acupuncture-treated groups. A number of venerable fertility centers now formally incorporate acupuncture as an integral program. Furthermore, there are no discernible side effects.

Some may mention that there is no scientific basis for these treatments, oblivious to the fact that, besides the treatment for infections, many therapies are empirical and their mechanisms remain unclear. As for the patients, they are less concerned with how the treatments are accomplished as long as their cure is successfully achieved.

Dr. Li Zheng is a well-respected acupuncturist and herbalist based in Boston. As an active clinician, as well as a Ph.D. researcher, she has written this book to demonstrate the rationales and procedures of these acupuncture-based fertility treatments. We celebrate her endeavor and expect grateful readers to benefit immensely from this marvelous book.

Shiao-Yu Lee, M.D.
Faculty Emeritus
Obstetrics Gynecology and Reproductive Biology
Harvard Medical School

Contents

Introduction to Infertility

Hypothalamus- pituitary-ovary axis:

The hypothalamus is located at the base of the brain. It controls our thirst, hunger, sex drive, body temperature, sleep, mood and the release of hormones from a master gland called the pituitary gland. The pituitary is a small oval gland in the brain producing hormones that control the thyroid, adrenal glands and ovaries. It also produces growth hormones, antidiuretic hormone, prolactin, and oxytocin.

In order to achieve a healthy pregnancy, the hypothalamus-pituitary-ovary axis has to be in balance:

Before the ovarian follicles are signaled by the follicle-stimulating hormone (FSH) from the pituitary, the follicles respond to hormone-regulating factors within the ovary itself. One important factor that emerges to influence ovarian health is called insulin-like growth factor. Other hormonal factors in the ovaries are: insulin-like growth factor binding protein, interleukin, tumor necrosis factor, inhibin, vascular endothelial growth factor, and activin. These ovarian growth factors help determine the eventual fertility potential of the egg. So it is very important to maintain healthy ovaries with all of these factors in balance a couple of months before attempting to get pregnant.

Every month, the hypothalamus secretes gonadotropin-releasing hormones, which stimulate the pituitary to produce follicle-stimulating hormone (FSH). FSH stimulates the ovaries to produce estrogen and progesterone. The hypothalamus also secretes corticotrophin-releasing hormone, which stimulates the pituitary to produce more adrenocorticotropic hormone (ACTH), stimulating the adrenal gland to produce cortisol, progesterone, and other hormones. The ovaries produce most of the progesterone, however; only a small portion of progesterone comes from the adrenal glands. Ovary and adrenal gland function must be optimized.

The hypothalamus can also produce endorphins that bind to receptors in the ovaries. If a person produces too many endorphins, it might make the person feel good, but can potentially cause a problem for pregnancy. Professional athletes very often have menstrual problems and infertility because of endorphins.

Acupuncture and Infertility:

The meridian system consists of 20 interconnecting meridians with more than 400 acupuncture points. These acupoints are located on specific areas near the surface of the body which demonstrate higher electrical conductance because of the presence of the higher density of gap junctions (special connections along cell borders to transmit iron and other intracellular signaling molecules). They act as converging points (or sinks) for electromagnetic fields. A higher metabolic rate, temperature, and calcium-ion concentration are also observed at these points. In principle, positive (anode) pulse stimulation of a point inhibits the organ function, whereas negative (cathode) pulse stimulation enhances that function. This forms the basis of electro-acupuncture.

Acupuncture has been approved to help optimize all hormone levels by working through the Hypothalamus-Pituitary-Ovary axis. Traditional Chinese Medicine (TCM) facilitates the body's own natural production of hormones while restoring the health and harmony of the entire system. It cannot be said that acupuncture simply increases certain hormones; it has a dual role to micro-adjust different kinds of hormones at different parts of the hypothalamus-pituitary-ovary axis (HPOA) and restore a subtle balance.

From the Western medical point of view, the influence of electrical acupuncture on the change in estrogen receptor expression in the brain may be one of the mechanisms to normalize the function of the HPOA axis. Chang R et. al published an article in *Fertil Steril* 2002, suggesting that acupuncture positively impacts opioid production in the central nervous system which, in turn, influences gonadotropin secretion from the hypothalamus.

Dr. Paul Magarelli, a reproductive endocrinologist and the medical director of the Reproductive Medicine & Fertility Center in Colorado, carried out clinical research combining acupuncture and in-vitro fertilization (IVF) procedures. Of those women in the acupuncture group, 51 percent achieved a successful pregnancy, compared to 36 percent of those who underwent IVF alone. Comparatively, the miscarriage rate was 8% and 20% in the acupuncture versus non-acupuncture group, respectively. There were no ectopic (tubal) pregnancies in the group of women who underwent acupuncture, but 9% of those in the group without acupuncture had a tubal pregnancy. They also reported 23% more births per pregnancy among those who had acupuncture as part of their infertility treatment. Initially, Dr. Magarelli did not believe in acupuncture. After this research, he was convinced that acupuncture does help to improve the IVF success rate.

Chapter One

Endometriosis and Fibroids

According to the National Institute of Health, 70-90% of women will suffer from uterine fibroids at some point, and an estimated 5 million have endometriosis, causing pain, infertility, lost work days, and over 200,000 hysterectomies each year in the United States. There are alternative therapies and pro-active measures women can take to prevent their occurrence. Hysterectomies can cause prolapse of the uterus, bladder and early onset of menopause symptoms.

What is endometriosis?

Endometriosis is a condition in which a piece of endometrium, uterus lining, grows outside the uterus such as ovaries, bowel, bladder, or fallopian tubes. The condition affects 10 to 15 percent of American women. Every month, when a woman has her period, this piece of endometrium outside the uterus starts bleeding and eventually forms a lot of adhesions and tubal blockage with severe menstrual pain.

Symptoms of endometriosis:

- Diarrhea and painful bowel movements, especially during menstruation
- Intestinal pain
- Painful intercourse
- Abdominal tenderness
- Severe cramping
- Backache
- Severe menstrual cramps
- Excessive menstrual bleeding
- Pain in the pelvic region with exercise
- Painful pelvic examination
- Infertility

What are Fibroids?

Fibroids are the abnormal growth of uterine muscle. Fibroids can vary greatly in size, shape, and location. They can grow inside the uterus, uterine wall, or on the surface of the uterus. They can be as big as a cantaloupe or as small as a sesame seed. They are usually benign, but if estrogen levels are too high along with uterine inflammation, fibroids can become malignant.

Symptoms of fibroids:

The symptoms include heavy bleeding or painful periods, bleeding between periods, feeling of fullness in the pelvic area (lower abdomen), frequent urination, pain during sex, lower back pain, infertility, having more than one miscarriage, and having early onset of labor during pregnancy.

Why do fibroids occur in younger women more frequently than before?

There are three main types of human estrogen: estrone, estradiol and estriol. They occur at a ratio of 15:15:70, according to Dr. Michael Lam, a famous physician specializing in nutritional and anti-aging medicine. Estradiol has the strongest effect on stimulating tissue growth. Estrone is the second strongest; the weakest is estriol, which has an anti-carcinogenic effect. Birth control pills have an estrogen ratio of 75:15:10. In this artificial combination, estrone has been increased from 15% to 75%, while estriol, the weakest and safest estrogen, has been reduced from 70% to 10%. Simply put, birth control pills dramatically increase the stimulating effect of estrogen compounds while decreasing the amount of beneficial estrogen, estriol, by 60%.

Birth control pills are designed to stop ovulation. Without ovulation, the production of progesterone drops tremendously. Because of the more intense estrogen stimulation with estrone from the birth control pill, the endometrium grows thicker with each cycle, but the period becomes very light and short. Consequently, during each cycle, the uterus cannot cleanse its internal membrane completely. Gradually, conditions such as fibroids and endometriosis can develop from the accumulation of the endometrium. Ironically, when young women are first diagnosed with fibroids and severe cramps, they are prescribed birth control pills. Although their cramps are initially reduced, and their periods much lighter, after five to ten years on the pill, most of these young women end up having surgery for their conditions. I have been tracking many young women who have taken birth control pills for over five years; they are not cured of either fibroids or endometriosis. Instead, they are often compelled to have a hysterectomy.

Birth control pills also influence the absorption of vitamins and minerals in women's bodies; this could also reduce the production of progesterone, leading to estrogen dominance. Estriol is called the "pregnancy estrogen" because it increases a thousand-fold during pregnancy. Very few women get cancer during pregnancy because of higher estriol levels, so it is said that this hormone is anti-carcinogenic. Because there are good and bad types of estrogen, women increase their bad estrogen levels and decrease their good one by taking birth control pills.

Why do endometriosis and fibroids cause infertility?

Endometriosis and fibroids change the condition of endometrium and negatively affect proper implantation, making the pregnancy very difficult. Endometriosis can cause scar tissue in the reproductive system, leading to adhesions or total blockage of the fallopian tubes. Hormone imbalance is the cause of these two conditions which can lead to infertility.

What causes fibroids and endometriosis?

If estrogen is not balanced with progesterone, it can stimulate the uterine tissue to grow abnormally, leading to fibroids. Clinical data shows that overweight women, typically with higher estrogen, tend to develop fibroids, indicating that an estrogen dominance condition can be related to fibroids.

When menstrual blood flows backward from the uterus into the fallopian tubes and spreads into the pelvic tissues, part of the uterine lining can also flow with the blood and implant outside the uterus. However, not all women who have menstrual back-flow develop endometriosis. There are many factors that may contribute to endometriosis and fibroids.

One factor is an imbalance of estrogen and progesterone; Khaleque Newaz Khan et al. in Japan published an article correlating estrogen and progesterone imbalance with endometriosis in *Human Reproduction,* 2005. He measured one of the growth factors called hepatocyte growth factor (HGF), a chemical stimulating tissue growth, in one of the immune cells called macrophages under different conditions. Women with endometriosis tended to have higher levels of this particular growth factor compared to the normal control. Furthermore, when macrophages were exposed to a higher levels of estrogen without progesterone, this growth factor level increased, and its gene expression was also enhanced. When the estrogen level was dampened with another chemical, or the progesterone level increased, this growth rate became lower. A higher growth factor may facilitate the endometrium tissue growing outside the uterus when there is back flow of the menses.

Genetic makeup can be a factor. Some women have a higher base level of estrogen and tend to have a very thick endometrium build-up before their periods as well as very heavy blood flow and painful menstruation. People who eat an unbalanced diet and wear tampons tend to develop endometriosis when the menstrual blood flows backward. If a woman thinks she belongs to this group, she needs to eat more cruciferous vegetables such as cauliflower, broccoli, and cabbage to reduce her estrogen dominance condition. She can also take herbs to help her ovaries produce more progesterone. If a woman practices yoga, meditation and qigong, her stress level will be lowered, and her progesterone level will increase to balance the stimulating effects of estrogen. People who pray daily also tend to have a healthier life because they have more

7

relaxation hormones and less stress hormones. Rose Kennedy, President John F. Kennedy's mother, is a good example of using prayer to reduce stress: her faith helped her live many years with dialysis despite so much hardship.

What can improve fibroids and endometriosis?

1. Surgery can quickly remove the fibroids or scrape off the endometriosis in the uterus or outside the uterus. If a woman uses acupuncture and Chinese herbs to balance her hormones and immune system post-surgery, she can prevent the fibroids or endometriosis from recurring, and get pregnant faster. In my practice, I always follow up with patients who currently have fibroids or endometriosis, or who had surgery for this condition. However, if hormone balance cannot be restored after surgery, the fibroids and endometriosis will grow back very quickly. During menopause, these two conditions can lead to heavy bleeding, and women with fibroids and endometriosis are most likely to end up with a hysterectomy.

2. In Chinese medicine, fibroids and endometriosis are associated with stagnant blood and a phlegm condition. Women should eat foods that move blood stasis like kelp, lemons, limes, onions, Irish moss, a seaweed called bladder wrack, and linseed oil which contains omega-3, a natural blood thinner.

3. Fresh ginger not only helps strengthen the digestive system but is also another natural blood thinner that reduces inflammation and facilitates blood circulation.

4. Another natural blood thinner is MSM, the abbreviation for methyl-sulfonyl-methane, a special biological sulfur found in fresh fruits, vegetables and meats. If a person eats the recommended amount of fresh fruits and vegetables, their blood circulation will improve.

5. An acupuncturist can prescribe special Chinese herbal formulas used to treat fibroids such as Tao Hong Si Wu Tang with Xiang Fu, Lu Lu Tong, Shi Chang Pu, Yi Yi Ren, taken for three months to one year.

6. A very common treatment in China is micro-electrical current therapy by cooking the Chinese herbs: Tao Ren, Zao Jiao Ci, Bai Jiang Cao, Hong Teng to make a concentrated solution, soaking gauze with the

decoction, and applying electrodes on the gauze attached to the uterus and ovarian areas over the abdomen. The micro-current will help the active ingredients in the herbal decoction to infiltrate the uterus and ovary areas.

7. Enemas with 100-200ccs of herbal decoctions, two to three times a week are very effective to treat tubal blockages, fibroids and endometriosis. It is a routine treatment for infertility in the hospital where I was trained in Beijing.

8. Avoiding tampon use because the back flow of blood is one of the factors causing endometriosis.

9. Avoiding the consumption of too many spicy foods which increase internal heat and inflammation in the uterine lining. Chronic inflammation can stimulate the tissues to grow abnormally and may change the gene expression.

10. Eliminating simple sugars from the diet to help restore the estrogen and progesterone balance. High sugar can lead to high insulin, which is growth factor itself.

11. Avoiding alcohol after ovulation, because progesterone levels rise post-ovulation, causing more internal heat

and phlegm. Alcohol can also enhance aromatase which facilitates the production of estradiol, causing an estrogen dominance condition.

12. Drinking more water especially during menstruation; this will help reduce inflammation and help liver and kidney function.

13. Avoiding the use of plastic containers because the chemicals can mimic the bad estrogen effect, stimulating the tissue to grow more.

14. Electrical acupuncture can not only improve the ovarian function to produce more progesterone but can also cleanse the uterus to get rid of old blood and grow the new uterine lining during menses. Patients will have shortened spotting time and less chance of getting an infection or inflammation in their uterine lining. The best time to cleanse is when a woman has her period or is still spotting. If the period is too heavy, one should wait until the period becomes less heavy.

Case One: Endometriosis and Infertility

Margaret was a 40-year-old with a pretty healthy life style. She drank one cup of coffee every morning, ate a healthy diet, and slept very well. Her periods were heavy the first day with some clots and cramping. She was on birth control pills (BCP) for three years. She tried to conceive for almost a year before starting IVF. One month before IVF treatment, she came to see Boston Chinese Acupuncturists. During the IVF cycle, she did not have as much discomfort as others, just a light frontal headache.

She only produced three follicles, and two fertilized eggs were put into her uterus. The whole procedure seemed very easy to her. She did not have crazy dreams, only a yeast infection while she was using progesterone which went away once she stopped it. She was one of the lucky women who got pregnant with the first round of IVF.

She stopped acupuncture when she was 12 weeks pregnant. During her emergency C-section for her first birth, she had two hours of bleeding. The doctor also discovered she had a lot of adhesions around the uterus which may have been from endometriosis. Since she followed a very healthy diet, her endometriosis did not cause severe discomfort. But her endometriosis did cause infertility due to the adhesions in the fallopian tubes.

In this case, she had responded well to hormone treatment because the hormone drugs were stimulating her Yang energy, and her Chinese medical diagnosis was spleen and kidney Yang deficiency, which was confirmed when she did not have discomfort from the Yang-stimulating drugs. The most important thing is that her Yin, blood, and other nourishing fluids were not deficient due to her very healthy lifestyle.

When she returned to acupuncture for her second pregnancy, her joints hurt every morning. She had extreme fatigue with a heavy sensation all over her body, even with 8 to 10 hours of sleep a night. She developed migraine headaches with vision problems before her periods. I strongly recommended that she take herbs to improve her ovarian function. After two months of taking herbs, her joint aches and migraines were gone, but she was also taking the fertility drug Clomid which challenged her FSH level, at a low of 7.9.

She was qualified for another IVF. This time was different. She had crazy dreams in which she was flying with a horse after she started her Lupron injections. She only produced one big follicle and two very small follicles, and her estradiol was relatively low. After two days, ultrasound showed that the big follicle had shrunk while the two small follicles had disappeared. Her estradiol went down quickly. Her IVF cycle was cancelled because there were no follicles to transfer.

After another two months of acupuncture and herbal treatment, she started a third IVF cycle. Her FSH was even lower this time: 5.8. The endocrinologist changed her protocol with BCP to suppress her ovarian function at the beginning. She stopped herbs five days before her injection started, two years after her first IVF. This time she developed five good follicles, and three were fertilized, although her estradiol was a very low 139; it should have been 500.

Before the physician decided whether to transfer the embryo, my patient asked how acupuncture could bring the estradiol up; otherwise, the procedure might be cancelled again. I did acupuncture three times a week and suggested that she use food therapy with yams since she did not want to take Chinese herbs during the procedure. After a couple of days, her estradiol was up to the required level. After the transfer, she started taking estrogen and baby aspirin with an injection of progesterone. She developed a burning sensation caused by a yeast infection with artificial progesterone.

Again, the third IVF failed. The infertility specialist explained that it may be due to the swelling of the fallopian tubes. For the fourth IVF, she had a brownish discharge the whole time she was on BCP, then she got her period earlier than usual. Furthermore, BCP caused severe mood swings, enlargement of the veins on her lower legs, and insomnia. This time she produced six eggs, three were fertilized and transferred. Hormone levels seemed normal. Ultrasound indicated no tubal swelling. Unfortunately, she did not get pregnant. No reason was given to explain this fourth failed IVF.

Now three years after her first successful IVF, she started her fifth round of IVF. Luckily, her FSH was still pretty low, indicating her ovarian reserve was good. Her thyroid looked slightly enlarged, and she could not warm up for a long time after exposure to cold windy weather, suggesting a low thyroid function. Interestingly, her pelvic exam showed no adhesions, and her endometriosis was gone after a year of acupuncture treatment. Because of a maintenance issue with the facility, her IVF procedure was delayed an additional five days, and she was on BCP 15 days with continuous spotting. She got her period on day 19 and only produced three follicles.

In this case, after two hours of severe bleeding during the delivery of her first child, her blood was shunted to the vital organs: brain and heart. Her ovaries did not get enough blood flow for a while because she was not checked on after delivery. Because of this, her ovarian function was severely compromised, plus she continued using BCP to suppress the ovarian function.

She not only had Yang but also Yin deficiency. With all the Yang drug stimulation, her Yin was even more depleted. Her estradiol dropped earlier during the procedure, and her implantation could not continue due to compromised ovarian function, even though her lining showed normal thickness and her hormone levels stayed within normal range. This is an example which shows FSH is not the only indicator of optimized ovarian function. But with the acupuncture and herbal treatment, her endometriosis and adhesions were gone.

Case Two: Endometriosis and Green Juice

Beth had been working for a computer company in a high-stress atmosphere. She was a very thin woman who loved her coffee and sweets. She had been on BCP for many years to relieve her menstrual pain due to endometriosis. She tried a few times to get rid of the BCP before she got married, but the menstrual pain and acne got so bad that she had to resume taking BCP.

When she came to see Boston Chinese Acupuncture, she was still taking BCP but planning to get pregnant within a few months. She was so worried about her menstrual pain that she had to come to the acupuncture clinic seeking help.

Her acupuncturist recommended that she reduce her coffee intake to stabilize her insulin levels and reduce stress. In the meantime, she cut down on sweets and started drinking a green juice with kale, spinach, lemon, and other fruits. She also had acupuncture treatments once a week, and stopped her BCP. She was surprised that her acne did not come back, and her menstrual pain was reduced. She was thrilled and continued with acupuncture treatments for three months. Then she called to say that she had become pregnant during the fourth cycle after she started acupuncture treatment.

In this case, the kale and other green vegetables helped stabilized her insulin levels, and her ovaries were able to produce more progesterone with sufficient minerals and vitamins. The acupuncture treatments also helped reduce her stress level, so her adrenal glands produced more progesterone. With sufficient progesterone, her uterus cleansed more thoroughly with each menstrual cycle. She had less testosterone to stimulate her hair follicles which got rid of her acne. She delivered a healthy boy at the end of 39 weeks.

Chapter Two

Ovarian Hyper-Stimulation Syndrome

When using fertility drugs to stimulate the ovaries to produce more than one egg, ovaries can be over-stimulated, producing more than 10 to 20 eggs at the same time with extreme swelling, abdominal distention and water accumulation in the abdominal cavity.

Symptoms of ovarian hyper-stimulation syndrome:

They are: ovarian enlargement, fluids leaking out of blood vessels and accumulating in the abdominal and chest cavity, electrolyte imbalance, thrombosis, pain, nausea, vomiting, diarrhea, abdominal distention, difficulty breathing, decreased urination, kidney and liver dysfunction.

Who gets this syndrome?

1. Patients with polycystic ovarian syndrome.
2. Patients with hypothyroidism. The increased thyroid stimulating hormone (TSH) stimulates the FSH receptor, so the body over-produces HCG.

Because excess follicles develop at the same time, eggs cannot mature properly and the corpus luteum function may be insufficient, leading to an uneasy pregnancy after ovulation. If the corpus luteum does not produce enough progesterone, it will negatively influence the development of the fetus and cause a possible miscarriage.

Fortunately, acupuncture and Chinese herbal medicine can help reduce the symptoms and induce the remission of ovarian hyper-stimulation during in vitro fertilization (IVF). Acupuncture can strengthen the kidney and lung function to facilitate the absorption of abdominal and chest fluids by the body to prevent kidney failure.

Case One: Ovarian hyper-stimulation and hypothyroid

Fran had a beautiful daughter who was breast-fed for more than one year. She had been trying to get pregnant for a couple of months after the first pregnancy. The hormone panel check-up was normal. When Fran first came to see me, she had just finished two intrauterine inseminations (IUI) with Clomid. The next step was to try a stimulated IUI with the injection of hormones before trying an IVF cycle. Her symptoms clearly showed Chinese Medicine patterns of spleen deficiency, liver qi stagnation and blood stasis: heavy periods, clotting, cramping, yeast infections around ovulation, an itchy sensation around the genital area, loose stools, and fatigue with depression before her period. Other symptoms were migraines on the day before and the first day of the period, irregular heartbeat and hemorrhoids.

We started acupuncture treatments two months before her stimulated IUI. She took *Clomid* first which made her body temperature go up and down. On day three of her period, she injected FSH in preparation for the intrauterine insemination. After a couple of days of stimulation, she developed ovarian hyper-stimulation syndrome with a severe lower abdominal pressure sensation, heart palpitations, trembling, nausea, headache, edema, and carpal tunnel pain. The original IUI was changed to IVF due to ovarian hyper-stimulation syndrome.

On day 15 of her last period, four good embryos were transferred. Fluids had accumulated in her abdomen; she felt so bloated that she could hardly bend over to pick up her daughter. Ultra-sound showed extremely enlarged ovaries with water retention in the abdominal cavity. With acupuncture treatments, she did not develop lower leg swelling. Her kidney and spleen function were strengthened with acupuncture to help discharge the extra water.

On the 11th day after embryo transfer, her blood test showed that she was pregnant. After two weeks of acupuncture treatments twice each week, her hyper-stimulation symptoms were gone. I saw her again when she was 37 weeks pregnant. She came to see me because her baby was breached. I showed her husband how to do moxibustion on UB67, and she felt a very strong baby movement afterwards. But they did not use moxibustion long enough to turn the baby; instead they went to see a midwife for manipulation of the baby position. During the procedure, the baby's heart rate went down, and obstetrician had to do an emergency c-section. She delivered a healthy baby boy.

After the delivery, her thyroid function became abnormal, and she was put on *Synthroid* to adjust her thyroid stimulating hormone (TSH) level. Her physician had difficulty normalizing her TSH levels, and she was exhausted with this condition on top of caring for two children. She gave up her plan to have a third child.

In this case, she may have had a sub-clinical hypothyroid condition before she tried IVF with the fatigue and other symptoms which may be the reason that she developed ovarian hyper-stimulation syndrome. During pregnancy, her thyroid had to work extra hard to provide enough energy. With the elevated progesterone from her pregnancy, her thyroid could still function normally. After delivery, her progesterone and other hormones dipped, the sub-clinical hypothyroid condition became clinically significant without herbal or acupuncture support.

Case Two:

Ellen, a 31-year-old woman, had started IVF. Her reproductive endocrinologist harvested 28 eggs. Two days later, she had three fertilized eggs transferred into her uterus. One day after the transfer, she started injections of progesterone. Five days after beginning those injections, she showed the following symptoms: abdominal distention, bloating, weight gain, nausea, irritability, insomnia, fatigue, shortness of breath and decreased urine output.

She was diagnosed with ovarian hyper-stimulation syndrome. Her abdomen became enlarged with water seeping out of her blood vessels. On days 9, 10 and 11 after egg transfer, she had a total of 1500 mL of water taken from her abdominal cavity. She had an infusion of albumin, 100-150mL, and other medications through IV every day to reduce the fluid seepage. After two weeks, her *BUN*(kidney function indicator) increased, suggesting her kidney function was compromised.

Her tongue had normal color and a thin white coating. Her pulse was wiry and soft. After trying all the Western medications, she was referred to Boston Chinese Acupuncturist and Herbalist to treat her deteriorating kidney function. The Traditional Chinese Medical (TCM) diagnosis was water retention due to spleen and kidney Qi deficiency. The herbalist gave her a formula, called Quan Sheng Bai Zhu San with Wu Ling San. The following are the herbs in the combined formula: Gui Zhi 6g, Fu Ling

10g, Ze Xie 10g, Bai Zhu 15g, Da Fu Pi 10g, Chen Pi 10g, Tong Cao 10g, Che Qian Zi 10g, Sheng Jiang Pi 10g.

After one dose, her urine output increased. After three doses, her abdominal distention decreased. Then, the herbalist stopped her herbs; Ellen continued to stay in the hospital. Four weeks after the egg transfer, she had abdominal pain and spotting, but her pregnancy test was positive. At that time, she still had some abdominal water, but her urination was normal. Her tongue was slightly pale with thin white coating. Her pulse was soft. The TCM diagnosis was Chong and Ren meridian deficiency unable to nourish the fetus. The following raw herbal formula was prescribed to her: Dang Shen 15g, Huang Qi 15g, Bai Zhu 30g, Shan Yao 15g, Sang Ji Sheng 15g, Shu Di Huang 30g, Tu Si Zi 24g, Shan Zhu Yu 15g, Gou Qi Zi 15g, Xu Duan 15g, Du Zhong 9g, Bai Shao 8g, Er Jiao 10g, Chen Pi 10g, Sha Ren 10g, Fu Ling 10g, Gan Cao 3g.

Chapter Three

Imbalanced Immune Function

When women get pregnant, the immune function must be somewhat suppressed to allow the fetus grow inside the uterus. If a woman's immune function is not balanced, their bodies will attack the fetus in the very early stages of pregnancy, causing unsuccessful implantation, miscarriage, or, in some cases, the fetus simply stops growing with a decreasing heart beat. If the woman has an autoimmune disease, the immune system is out of balance and she is very likely to have different problems during pregnancy.

What can tip the balance in the immune system?

1. Estrogen dominance: estrogen is not balanced with progesterone. When a woman is pregnant, increased progesterone can suppress her immune function to a certain level. That is one of the reasons that

allergies get better during a healthy pregnancy. Birth control pills (BCP), dysfunction of the ovaries, and too much stress can all cause an estrogen dominance condition.

2. Adrenal fatigue or insufficiency: it was already mentioned that cortisol helps to suppress the immune function. If cortisol is too low due to depleted adrenal function, women can produce antibodies that attack the fetus. Caffeine, high stress, and lack of sleep are known to drain the adrenal function.

3. High prolactin can sensitize the immune system: High stress, tumors in pituitary gland and a prolonged period of breast-feeding can lead to high prolactin. The most recent studies done by a Chinese scientist Liu, XY and published in Zhong Guo Zhong Xi Yi Jie He Za Zhi, showed that electrical acupuncture can lower the prolactin level in rats and increase the pregnancy rate. Chinese Herbs like malt barley sprout, Bai Zi Ren, Gou Teng, He Huan Pi can lower the prolactin levels, which may be associated with lowering stress levels, and have a calming effect.

4. Imbalance of the sympathetic and parasympathetic nervous system: in the autonomic nervous system chapter of *Acupuncture and Hormone Balance*, there was discussion how the sympathetic nervous system enhances the immune function, while the parasympathetic nervous system can suppress the immune function. If one drinks too much coffee, the sympathetic nervous system could be overacting. Patients tend to develop allergies and autoimmune diseases such as MS and rheumatoid arthritis.

5. Too much alcohol can influence our immune function due to a changed chemical balance.

Case

Lydia, a 39-year-old woman with a school-age daughter, had been trying to conceive for the past 5 years. She tried IUI twice before and was not successful. She also had developed an ovarian cyst during the procedure, so she decided not to repeat the IUI or IVF procedures. When Lydia came to see me, her hormone checkup was normal. I suspected there was a sub-clinical imbalance of her immune function. Several symptoms indicated that her immune function was out of balance. Her tonsils were removed when she was a little girl. She developed seasonal allergies with nasal congestion and runny nose in her early twenties. If she ate fruit skin, she developed itching in her throat and ears. Garlic caused her to become bloated with cramps and diarrhea, especially before her period.

Why did she have an imbalanced immune function? There were two problems. First, her spleen function was weak. She bruised very easily and could not digest even fruit skin. She also had a fungal infection on a fingernail and eczema on her hand. Also, she had bloating, gas, and sometimes diarrhea before her period.

The second factor was Liver Qi stagnation with blocked energy flow in the upper torso. Her face looked reddish, and she had chapped lips. She developed canker sores before her period when her progesterone and estrogen levels went up. She had breast tenderness, irritability, and felt warm during the night before her period. During her period, she had severe abdominal cramps. She tended to think a lot and talked very fast, and she became anxious even when there was no emergency at that moment. She was very thin, even though she had a good appetite and ate enough. Her temperature chart showed an irregular pattern. She ate a lot of sweets and drank many cups of coffee. In order to balance her hormones, she stopped eating non-nutritious food and coffee. She even stopped working outside the home to reduce her stress level.

She was highly motivated to get pregnant. After she had her first child and breastfed for a year, her period became scantier compared to her period before the first pregnancy. The eczema she had developed as a child now occasionally flared up just before her period. Her husband had a normal sperm count, motility and morphology. Her tongue was slightly red-tipped, with a thin white coating, and her pulse was thin and wiry. When she first came to see her Chinese acupuncturists, she had an irregular period from day 25 to day 32 with tender breasts, cramps, nausea, and irritability. After the first pregnancy, she developed an urgency to urinate and sometimes could not hold her urine.

With all her allergic symptoms, her immune function and her digestion were not optimal. Her fertility temperature chart showed a zigzag pattern. After a month of acupuncture treatment, her PMS symptoms were much better. She did not have cramps before or during her period. During her next cycle, her temperature was more stable after ovulation, but still not high enough.

She asked if she could start herbal powders; she faithfully took different formulas before and after ovulation. Her cervical fluids were very dry after ovulation before starting acupuncture and herbal treatment. After two months of herbal intake, her vaginal fluids increased, and her menstrual blood flow increased during her period, spotting before and after her period stopped.

But her body temperature still was not high enough for the entire luteal phase. Her energy increased with herbs. After four months of herbal formulas, her luteal phase increased from 10 days to 14 days, but still her temperature dropped in the middle of the luteal phase. After recommending that she take raw herbs, Lydia cooked and took a raw herb decoction for two months.

Finally, her temperature chart became biphasic. She stopped developing canker sore before or during her period. Her PMS symptoms were far less severe. Her anxiety lessened; she felt more relaxed and happy after taking raw herbs for four months. With her increased hormone levels, her temperature chart showed biphase, but she developed occasional headaches which lasted for shorter period of time. She also had mood swings just before her period from the sudden drop in her hormone levels. Five months after she took raw herbs, she had a late period; she had a lot of virginal discharge, indicating she may have had an early implantation.

After nine months of acupuncture and six months of herbal intake, she got pregnant naturally. She also got rid of her severe headaches, allergies, eczema and canker sores. Just before she found out about her pregnancy, on day 27 of her last period, she started to feel dizzy with vomiting and severe breast tenderness. On day 34, her pregnancy test was positive. Her first HCG was 290.

Interestingly, her allergy to certain fruit skins came back once she got pregnant. If she ate pineapple, her tongue became itchy and she would develop a runny nose and phlegm would build up in her throat. This is another example of how hormones can influence our immune function.

This case clearly demonstrates how the digestive system (spleen and stomach function) is closely connected with the immune function. If non-nutritious foods are regularly eaten, of course the body will respond with allergies. Furthermore, a person's genetic makeup can also play a role in the body's responses to system imbalances.

Chapter Four

High FSH and Infertility

The pituitary gland secretes follicle stimulating hormone (FSH), stimulating the ovaries to produce eggs. If the ovaries do not function well, the pituitary will try to produce more FSH to push the ovaries to work harder. Young ovaries do not need much FSH to function normally. Generally, if the FSH level is above 10, ovarian function is compromised. If it is above 25, a woman may start menopause prematurely. But FSH can fluctuate significantly between ages of 30 to 45 before menopause. In one cycle, FSH can be 13; in the next it could be 7.

So if FSH is slightly high, there are many methods available to improve ovarian function, so that the pituitary and hypothalamus does not have to try so hard to stimulate the ovaries with high FSH. When estradiol is high due to an artificial hormone intake or estrogen dominance, FSH could be lowered. Even with a very low FSH, if other hormones are out of balance, it is still very difficult to get pregnant. Another hormone called inhibin B, produced by good follicles, can also be used to predict the ovarian reserve. A lower inhibin B level may indicate poor prognosis of IVF. The following is a good example showing one can improve ovarian function by lowering FSH levels.

Case One: High FSH

Paula is a 43-year-old newlywed who had a miscarriage. Her FSH was around 36 on day three following her miscarriage. She was told that she had already started menopause and that her only choice was to have a child using a donor's egg, but she did not want to use fertility drugs. Two months after her last miscarriage, she got her period. On her own, she bought some Chinese herbal pills from the supermarket to improve her fertility. Another blood test showed that her FSH was around 5 on day three of a Clomid Challenge test. The Clomid made her miserable; she had night sweats, mood swings, and breast tenderness.

She decided to come for acupuncture and herbal medicine since her FSH had dropped dramatically after she took herbal pills for a short period of time. She had unremarkable symptoms: her blood pressure was generally very low, she usually had very irregular periods with no clots or menstrual cramps, slight low back pain and low sex drive. She loved to eat spicy food. She was slightly overweight, and had polyps in her uterus. Her tongue was of normal color, and her sublingual vein was slightly purple, indicating poor circulation. Her pulse was thin and small, which did not match with her constitution possibly suggesting poor circulation and a phlegm condition.

All her symptoms pointed to compromised ovarian function, but she still had the potential to improve the function because her FSH did drop after she took some herbal pills for a month. Since the pills were not an exact fit for her situation, she developed dry mouth, night sweats and irritability. After only six acupuncture and herbal treatments, her low back pain, dry mouth and night sweets were gone. Her body temperature stayed around 99 degrees after ovulation and held for the whole luteal phase.

In the next cycle, she discharged a lot of clots and her sublingual vein was less purple, indicating that her blood stagnation was improving. She was asked to cut down on spicy food intake and eat more regularly instead of just three big meals.

During her third cycle, she showed a clear discharge on day 18 and 19, and she had intercourse on day 13, 15, 17, 18 and 23, however, her temperature did not stay up high enough. It dipped on day 17 then again on day 22. During her fourth cycle, she continued herbs, and her period did not come until day 40. By day 50 of this period, the pregnancy test was positive. During the 8th week of pregnancy, she started spotting. She asked for herbs to prevent a miscarriage from her Boston Chinese acupuncturists, which were prescribed to strengthen her spleen and kidney function.

Fortunately, she only spotted one time. She continued herbs for two weeks. After her HCG went to 25000, there was no need to take herbs any more. Her reproductive system was ready to support the pregnancy. She ended up developing gestational diabetes with high blood pressure and severe leg swelling. She had to stay at home for the last two months and had a scheduled C-section with a healthy baby boy.

In this case, Paula's spicy food caused a phlegm condition and dried her valuable kidney Yin which naturally starts decreasing after 40, so she had internal heat with blood stasis. Since her FSH fluctuated so much, there was potential to improve her ovarian function. The herbs helped clear up the internal heat and restore the kidney Yin, so within four months, her ovarian function improved dramatically. If she had not changed her eating habits or taken the Chinese herbs, but instead jumped directly into IVF, she would not have gotten pregnant; and she may have started menopause earlier without sufficient kidney Yin to balance all the Yang fertility drugs.

Not every woman with high FSH can get pregnant with herbs within three to four cycles, but Chinese herbal formulas definitely can improve hypothalamus, pituitary and ovarian functions. Perhaps if Paula had taken herbs for another couple of months, her body may have had better control of her blood sugar levels and blood pressure. Her excess liver Yang could have overacted on the Kidney Yin deficiency, causing high blood pressure and kidney failure, which could have been treated with a bitter-tasting formula.

Chapter Five

Improving Sperm Quality

Normal sperm profile

Total sperm count should be twenty million or more in one milliliter of semen. Sixty percent of sperm should have normal shape and structure with an oval head and a long tail. Large, small, tapered or crooked heads or kinky, curled or double tails are abnormal. Clinically, forty percent of sperm with a normal shape should be enough for having children.

Normal motility will make sperm wriggle and swim the last inches in order to reach and penetrate the egg. To produce offspring, more than 40 percent of the sperm should have normal motility.

What can be done to improve sperm quality?

1. Maintain a healthy body weight. Too much or too little fat will influence hormone levels and thus sperm quality. Men also need cholesterol to produce testosterone, so avoid taking statins.

2. Regularly exercise to improve sperm quality, but overexertion or exhaustion can compromise the sperm quality with increased physical stress and higher levels of cortisol.

3. Reduce stress levels to optimize testosterone levels.

4. Avoid eating too much sweet or spicy food to produce semen in a neutral pH environment, from 7.2 to 7.8. If it is too acidic, the sperm can be deformed, or there may be inflammation in the prostate, because the prostate gland produces acidic fluids while the seminal gland produces slightly alkaline fluids. These two fluids are the main part of semen.

5. Drink plenty of water: if a man is constantly dehydrated, his semen will become very thick, and the sperm cannot swim fast enough to reach the eggs, which can only survive for 1 to 2 days;

healthy sperm can stay alive for more than five days.

6. Avoid anti-histamine medications. These medications will decrease the secretions from many glands, including the prostate and seminal gland.

7. Avoid alcohol, which can also dry up the semen.

8. Avoid ibuprofen. According to a study published in December 2017 by the Proceedings of the National Academy of Sciences of the United States of America, men who took a daily dose of only 600 mg of ibuprofen had "compensated hypogonadism", resulting in a lowered testicular response to elevated LH hormones with gene expression changes in the testes.

9. Wear loose underwear to keep the temperature low around the testicles. Sperm temperature needs to be one to two degrees cooler than normal body temperature, which is only achieved when the testicles are not held too closely to the body by tight clothing.

10. Take multivitamins to provide antioxidant protection as well as trace minerals such as selenium and zinc, vital for healthy sperm.

In the past, there was less environmental pollution and less stress. With all the environmental and social changes, men have to take better care of themselves to improve their sperm.

Acupuncture and sperm quality:

Acupuncture has been proven as one of the most effective treatments to improve sperm quality with little or no side effects.

A study done by Per J. et al. in Germany in 2005, showed that acupuncture twice a week for five weeks increased the total number of sperm and the percentage of normal sperm in males who had unknown infertility.

In another study done by Dr. Siterman S et al. and published in Archeology Andrology in 1997, sixteen men with fertility problems were treated with acupuncture twice a week for five weeks. Semen samples were analyzed before and one month after treatment. When the semen analysis was compared to those from normal control, total functional sperm percentage, percentage of viability and total motile sperm number per ejaculate were significantly higher. The integrity of the sperm tail was also improved. The normal structure of the sperm tails and sperm motility were highly correlated.

Even if a man has a normal percentage of healthy sperm, with acupuncture treatments, his healthy sperm can be as high as 98%, and the motility can be over 80%. This happens to many of my patients who come with their wives to have acupuncture together, almost doubling their success rate for infertility.

Case Study:

Bob is a considerate husband who wanted to improve his sperm quality. Having lived for a time in Asia, he understood how much Chinese medicine could help. He took herbs and had acupuncture treatments once a week to improve his sperm quality. Although he was very energetic, he had anemia with an unknown cause; his fingernails were pale and flat, and he had occasional night sweats and easily sweated after light exercise. At the beginning of treatment, his total number of sperm was less than 20 million. The blood deficiency or anemia was probably caused by imbalanced digestive function. One possible reason he developed these symptoms is from his constitution or genetic make-up (kidney yin deficiency); the second may be associated with stress and over-thinking from his demanding job.

Three months after beginning treatment, his sperm count increased to 28 million, but the motility and the percentage of normal sperm were still less than 40%. After another three months of acupuncture treatments and herbs, his night sweats were gone, and the total sperm count per ejaculation increased from 28 million to 40 million. Furthermore, his sperm motility and normal sperm percentage increased to 60%. Later, his wife delivered a healthy baby boy.

Chapter Six

Hypothyroid and Infertility

In hypothyroidism, the thyroid gland is under functioning and not producing thyroid hormones necessary for metabolism, temperature regulation, an even heart rate. Some symptoms of a hypothyroid are feeling cold and tired, dry skin, poor circulation, brain fog, constipation, depression, and weight gain.

If the thyroid does not function well, the pituitary gland produces more thyroid stimulating hormone, which is associated with a high rate of unsuccessful pregnancies and miscarriages, according to a study by R. Negro, in Italy in 2007.

Hypothyroid conditions can also influence ovulation, and a higher percentage of women with hypothyroid show abnormal ovulation.

Improving thyroid function:

1. Eat foods that nurture thyroid function: raw greens, seaweed, walnuts, seeds, sprouted beans, and fresh vegetable juice.

2. Avoid the following foods until your thyroid function becomes normal: mustard greens, kale, peanuts, soy flour, Brussel sprouts, and unsaturated oils like flax, soybean, safflower, corn and canola oils. In Chinese medicine, soybeans are difficult to digest. Eating too much soy and soy-based products can cause a phlegm condition, which may aggravate a hypothyroid condition. If consuming flax seed oil causes the thyroid function to progressively worsen, then stop taking flax seed oil for a while.

3. Avoid adding sugar to coffee or eating too many sweets. High sugar induces high insulin, causing inflammation in many glands, including the thyroid. Also, high sugar can cause malabsorption of many minerals critical for thyroid function. Just like alcohol and drugs, sweets disrupt metabolic balance, causing dysfunction of many glands.

4. Eat fresh fruits instead of drinking juice. Surprisingly, sugar is added to juice, sometimes more than 14 teaspoons for one serving.

5. Add a small amount of seaweed to soups, providing enough minerals for hormone production.

6. Take herbs or medications to get rid of viruses and infections faster, which prevents them from becoming chronic. Chronic infections lead to dysfunctions of the thyroid gland and other glands.

7. Do activities to de-stress like acupuncture, yoga, Tai chi, and meditation. For women especially, balancing estrogen and progesterone with acupuncture and herbs will help to avoid hypothyroid and other chemical imbalances.

During pregnancy, a woman's adrenal gland and thyroid become depleted. If there is an underlying hypothyroid condition before pregnancy, it is better to treat it as soon as possible, which will help the pregnancy. Also, after delivery, the thyroid will have a better chance of functioning normally with treatment.

Case Study:

Tanya is a 40-year-old female. She became pregnant easily the first time. After her first pregnancy, she developed a hypothyroid condition and began taking Synthroid. She experienced gas and abdominal pain after eating salads, she felt depressed, and her hands and feet were cold. When she first began acupuncture treatments, she had just had a miscarriage. Her regular checkup had been normal until she reached 12 weeks and found out that her fetus's heart had stopped beating. After a D & C, testing confirmed that the fetus had a chromosome problem. This began her long journey of trying to get pregnant with modern techniques.

Tanya's FSH was 13.9, and her blood pressure tended to be low. First, she tried two stimulated intrauterine inseminations (IUIs), and she wasn't able to get pregnant. The last IUI had brought her a "chemical pregnancy" with tissue growing in her body instead of a fetus growing in the uterus which kept growing and producing HCG. In order to kill this growing tissue, she had to take chemotherapy drugs. It took two and a half months to clear up this pseudo-pregnancy. Finally, she discharged a lot of tissue and gained 10 pounds by the end of this process.

Then, she started IVF. First, she took BCP to suppress her ovary function. During the first IVF, she was told that her husband's sperm had problems. She was convinced that an embryo biopsy could be used to make sure that the embryo did not have a genetic defect. Her ovaries only produced three follicles, which did not survive the embryo biopsy process called PGD. So, unfortunately her first IVF was cancelled after all the injections of hormones into her body. She felt very warm with cold hands and feet afterwards.

The next month, she received a second IVF procedure. Six eggs were produced and fertilized, but the chromosomes were abnormal. Her body also produced antibodies against her own thyroid. Although discouraged, she was determined to have a healthy baby. Her endocrinologist adjusted her medication for her thyroid problem.

She started the third IVF procedure with a new protocol, and her insomnia got worse with "brain fog" and extreme fatigue. This time, with six months of acupuncture treatments, she developed 19 follicles, and her hormone profile looked good. Although eight eggs fertilized, none survived the embryo biopsy procedure; the transfer was cancelled again. This experience took an emotional toll on her, but she is a tough and determined woman.

After these three unsuccessful IVF procedures, she found an endocrinologist in Las Vegas. After a two-month break, my tough patient flew to Las Vegas to try this new procedure and the most advanced techniques. The physical examination and hormone stimulation were done there; the transfer was done in New York. During an extensive examination, the physician discovered she had abnormal natural killer cells, which may have contributed to her infertility. She was given immune globulin (IgG) intravenously, took one of the glucocorticoid drugs to suppress her immune function and got an infusion of blood thinners to prevent blood clotting because she was taking birth control pills.

In this fourth IVF, she produced 16 eggs and got pregnant with this expensive new protocol. This time, at 5weeks she had a third miscarriage. During this IVF, the endocrinologist had frozen some eggs. While she was on BCP to get ready for the transfer of frozen eggs, her pregnancy test turned positive. Of course it could not be a healthy pregnancy due to the BCP. It took another two months to clean up this abnormal pregnancy before she was ready for the transfer of the frozen embryos. At this point, she had already tried four IVF procedures and had three unhealthy pregnancies. She had frozen embryos transferred. She had been on cortisone, blood thinners, antibiotics and Synthroid. Because of the suppression of her immune function, she got a leg infection and a throat fungus infection. Her heart rate increased so much that she could not sleep. The frozen eggs did not implant in her uterus even with the medications.

After another two-month break, Tanya tried the fifth IVF cycle. She produced 20 eggs. With many months of suppression of her immune function, her open wound on her left knee area could not heal for a long time. After only a couple of acupuncture treatments, her immune function got stronger and the open wound finally closed. With all the cancellation of her IVF cycles, she finally agreed that an embryo biopsy could not guarantee a healthy fetus, and it was too harsh for the delicate embryo. The three fertilized eggs from the fifth IVF were transferred without being tested for genetic defects.

After all her effort and more than a year of acupuncture treatments, she got pregnant; this time with twins. She felt cold and tired, and I knew she was pregnant. This time without the embryo biopsy, the two eggs happened to be good eggs. Her HCG jumped up from 400 to 900, then to 4000. She was so worried about a miscarriage that she kept checking her pregnancy hormone levels. Finally, at six and half weeks, the vaginal ultrasound showed two hearts beating and at her last check up, her pregnancy hormone was 25,000.

She was extremely tired. With all the hormone injections, she had many crazy dreams. She was transferred to see a high-risk pregnancy specialist who was strongly against cortisone during pregnancy. Also because of the high progesterone level, the blood thinner infusion may have caused life-threatening internal bleeding. While she visited this new specialist every two weeks, she continued with acupuncture treatments with me until 13 weeks.

Interestingly, with her hormone levels so high, her thyroid function improved. Her taste for foods changed. Before her pregnancy, she had to have pretzels every day. She was very addicted to salty foods. But after she got pregnant, she did not crave it at all. She had cravings for hamburgers, which she rarely ate before.

In this case, a hypothyroid condition played a very important role in her four miscarriages. IVF didn't really help her to get pregnant or prevent her miscarriages. This woman was able to get pregnant very easily even on BCP. Her primary problem was the quality of the egg and sperm. The three IUIs and five IVFs took more than two years; she still ended up with three abnormal pregnancies.

Fortunately, she strongly believed in acupuncture, which is a fraction of the cost of expensive IVF and other interventions. If she had taken herbs while doing acupuncture, it would have taken less time to get a healthy pregnancy. Plus, Chinese herbs would have helped to strengthen all the glandular functions. Even after this twin pregnancy, Tanya's thyroid function would take some time to recover fully, if at all. She told me that she would come back to get acupuncture treatment for facial rejuvenation. But I have to strengthen her adrenal and thyroid glands, otherwise she will lose her muscle tone quickly and have a very tough menopause with severe hot flushes, memory loss, night sweats and mood swings.

Chapter Seven

Recurrent Miscarriage

One of the most difficult life experiences is to suffer a miscarriage, yet many women do. According to the UK's National Health Services, between 15-20% of pregnancies will end in miscarriage before 20 weeks, with most of them occurring before 7 weeks. Yet only about 1 in 100 women will deal with recurrent miscarriage, which is losing three or more pregnancies in a row. Most of the time, there is an unknown diagnosis, yet there are underlying causes that prohibit a healthy pregnancy and ways to help overcome recurrent miscarriage.

What are some of the causes of miscarriage?

1. Hormone imbalance: Low progesterone will influence early implantation because progesterone supports the endometrium allowing the embryo to grow in the uterus.

2. Poor uterine lining: Low estrogen associated with poor ovarian function can lead to a very thin uterine lining. Anemia and inflammation are common factors causing dysfunction of ovaries and uterus.

3. Abnormal fetus: the sperm or egg can have abnormal chromosomes. Certain chromosome problems can cause very early miscarriages, while others can lead to late miscarriages.

4. Autoimmune disease: an imbalanced immune system produces antibodies which attack the sperm and fetus. Infertility specialist, Dr. Salim Daya of Ontario, Canada wrote in *Chatelaine Magazine* that 20 to 25% of miscarriages are associated with auto-antibodies.

5. Anti-histamine drugs: these drugs are known to cause fetal malformation which will eventually be discharged from the uterus.

6. Certain parasites found in cats can cause toxoplasmosis in humans, leading to miscarriage or even a deformed fetus.

7. Artificial sweeteners and MSG: too much MSG found in flavored chips, meat seasoning and packaged soup can be toxic to a pregnant woman.

8. Environmental toxins: chemicals such as pesticides can cause the body to produce more antibodies which attack the fetus according to a study at California State University led by Dr. Jack D et al.

9. Alcohol and cigarettes: a new Danish study finds that even moderate amounts of alcohol, five drinks per week, can increase a woman's risk of miscarriage, particularly during the first trimester. This study included 25,000 pregnancies from 1989 to 1996.

10. Too much coffee: a study of 1,909 women in Connecticut showed that the infertility rate increased by 55% in women drinking one cup of coffee per day, 100% higher in women drinking one

and one-half cups of coffee per day and 176% in women who drink 3 cups of coffee per day.

11. Exposure to certain chemicals: xylene, acetone and petroleum distillates. Women workers who used the above chemicals daily have a much higher miscarriage rates than normal.

12. Swimming in cold water can stimulate the contraction of uterus, causing miscarriage in the first trimester.

13. Staying in a hot tub too long can also harm early embryo development, leading to miscarriage.

How can miscarriage be prevented?

1. Avoiding alcohol and spicy foods can reduce internal heat, the underlying cause of inflammation in the endometrium of the uterus. One might wonder why people in generations past ate a lot of spicy food but could still carry a baby full-term. First, older generations did not have as much internal heat from stress, alcohol, too much meat

and other high calorie foods, so they needed a little bit of spicy food to help body produce more heat. Second, older generations usually had many pregnancies, and miscarriage was very common if they were pregnant more than four or five times. Now, people aim for healthy pregnancies for every attempt at conception.

2. Avoid cats. If there is a cat in the house, it should be restricted to a certain area. Cats carry a parasite that causes toxoplasmosis; this may cause miscarriage or abnormal development of fetus. Years ago, doctors advised pregnant women to avoid changing the cat litter so she could be safe during the pregnancy. This is not true. A microbiologist told me that cat feces can carry the parasite even when the feces are dry. If the cat sits anywhere, the parasites can be transmitted to the area when the cats move around the house. For the past 26 years, whenever a woman came to me with a history of more than two miscarriages, my first question was always, do you have a cat? Over 90% of women answered yes. One woman miscarried three times, and she had three cats roaming around the house. The fourth time she was pregnant, she did not tell any body until she

reached 16 weeks, during which the ultrasound indicated that the baby looked pretty healthy. She was thrilled to tell all the relatives and co-workers that she was pregnant. Unfortunately, the ultrasound showed the brain, limbs and spine were all deformed by the 20th week. Very few women love cats so much that they keep trying even with the cat walking around the house; eventually they may have a full-term pregnancy. The problem is miscarriage is very harmful for your reproductive system and mental health. Many women develop severe depression afterward. Furthermore, there is higher possibility that the baby can have some genetic problem.

3. If one has already had two or more miscarriages, acupuncture treatments once or twice a week should be started for the first 10 to 12 weeks of pregnancy. Acupuncture can relax the uterus and improve the ovarian function, so the balanced hormones will help prevent another miscarriage.

4. The following porridge can help prevent miscarriage: 1 cup of brown rice, 1 cup of Yi Yi

Ren "Job's tears", Shan Yao30g, adding 500 mL water, and cooking for half an hour.

5. Pressing the following points to prevent the miscarriage: CV6 (on the midline of the abdomen, the width of the index and middle fingers below the belly button), Kid16 (a pinky finger's width to the left and right of the belly button), Kid3 (on the ankle in the hollow between inner ankle bone and the Achilles tendon), PC6 (2 thumb-widths up the arm from the inner wrist-crease between the two tendons seen when the wrist is hyper extended).

6. Practicing meditation to reduce stress and improve progesterone level.

7. Avoiding hot tubs helps improve healthy implantation.

8. Taking herbal tea from an herbalist to balance hormones; the herbs for preventing miscarriage are very safe if given by a specialist in treating infertility.

9. Cooking this Chinese formula dispensed by an herbalist to clear internal heat and inflammation, strengthen the spleen function and support the fetal

growth in the uterus: Huang Qin6g, Sheng Bai Zhu10g, Fu Ling 12g, Shan Yao30g, He Shou Wu12g.

10. Avoiding swimming during the first 12 weeks of pregnancy to avoid uterine contractions.

Case One:

Gina was a 38-year woman with approximately 10 years of secondary infertility. For her first pregnancy, she tried for 5 years and had a healthy boy. She was never able to get pregnant again. She had intercourse approximately once per month, mid-cycle before she came to see her acupuncturists for acupuncture treatment. Her menses were regular, every 27 to 30 days, with occasional irregular period and no inter-menstrual bleeding. She had a history of oral herpes virus infection, but no history of sexually transmitted disease. Her ovulation tests were never positive, but an ultrasound showed that she did ovulate. Her FSH was 6.5 on day 10. Both of her fallopian tubes were open and her uterus was normal.

Her husband was a smoker, but his semen analysis was normal with 92 million sperm per mL with 74% motility, and 67% normal morphology. He had very bad allergies and often took anti-histamine drugs.

Gina's TSH was normal. She had mild depression and mood swings, and had recently started Effexor, Bactrim and Pyridium therapy. She had recurrent urinary tract infections. She looked lethargic with no facial expressions. She slept more than 8 hours but still felt tired. Her menses were dark red with clots. She had lower back pain and cramps before and during her period. A heating pad reduced her pain. Her legs were swollen for a long time and she always felt cold.

Since she had been trying for 10 years, I did not expect her to get pregnant for at least 6 months. She was very open to Chinese herbs and started taking herbs with acupuncture treatment right away. I also asked her to chart her body temperature. My Traditional Chinese Medical diagnosis was kidney and spleen yang deficiency, liver and heart fire. Her temperature showed no pattern at all. After only two and half months of acupuncture and herbal medicine, she had a positive pregnancy test. By day 37 of her last period, her temperature dipped, and she had slight cramps. She started spotting at 6 weeks and miscarried at 6 ½ weeks.

She had a sore throat, coughing, chest tightness, and nasal congestion during this pregnancy. Unfortunately, this time she miscarried.

After the miscarriage, she took some herbs for the sore throat and continued with acupuncture treatments and herbal medicine. Finally, her ovulation test turned positive. Her leg swelling was reduced, but her temperature was still highly variable. She had lower back pain before and during her period, indicating a kidney deficiency.

After another four months of treatment, she got pregnant again. This time at 6 ½ weeks, the fetus had no heart beat, but her pregnancy hormone HCG was normal. She had a D&C on the 8th week of this pregnancy. When I saw her after the D&C, her lips were swollen with many blisters, and she had a fever. I gave her some herbs to treat her herpes infection which cleared it up.

She stopped having acupuncture treatments while she had three IUI procedures at an infertility clinic, but she never could get pregnant. After trying for two years, she was convinced that her eggs were not good anymore and was ready to try IVF. With hormone injections, she produced seven follicles; two were fertilized and transferred, however she did not get pregnant with the IVF. After the procedure, she gained 10 pounds. Most of the weight went to her legs especially around the knee area. The swelling of her legs got worse in the evening and her weight fluctuated by five pounds a day. She had a massage and urinated so much that she instantly lost three pounds. She felt colder and more lethargic; she was slow to wake up in the morning after her IVF. Her period also got shorter.

Six months after her IVF treatment, she came back to me just for general health. Her main complaint was difficulty losing weight and water retention in her lower legs. She had a bug bite and developed rashes on her right lower leg which resulted in a large area of darker pigmentation. After each acupuncture treatment, the pigmentation on her leg became lighter. She still had a very hard time losing weight even though she did not eat very much. Later, I found out she ate a very late breakfast, lunch, and dinner which slowed down her metabolism by eating so late.

In this case, Gina's first miscarriage may have been associated with egg or sperm problems and hormone imbalances because the fetus was discharged spontaneously at 6 1/2 weeks. As for her second miscarriage, the hormones were pretty normal, but the fetus simply stopped growing with no heart beat at 8 weeks. There may have been many factors involved in her miscarriages, but one simple thing I suggested to her was to remove the four cats she had in her house. She told me that she did not clean the cat litters, but the four cats were free to roam everywhere. Nobody can guarantee that there would not be parasites in other places of the house. Unfortunately, Gina could not temporarily take the four cats away because her son would not allow it. Of course, her miscarriages could be due to her husband's sperm or her egg quality, but it is good to eliminate the known factors one by one. She already had some hormonal imbalances due to daily stress and her age. After all the hormone injections, her ovarian function was decreasing even more. Now she was 43 years old and had given up the hope of getting pregnant. The focus of treatment shifted to improving her general health and facial rejuvenation.

Case Two: Repetitive miscarriage

Cindy was a 40-year-old woman with a 4-year-old daughter who was very pretty and smart and in good general health.

In her early twenties, she developed a cyst in her spine and had surgery. Since then she had started acupuncture treatment periodically to strengthen her back. During this time, she tried for a year to get pregnant with a high prolactin level, a chemical stimulating the breast to produce more milk. She tended to have phlegm in her throat with the sensation to clear her throat.

When she started treatments, she had just miscarried a couple of weeks prior. Her main symptoms were insomnia, waking up in the middle of night and difficulty falling asleep, night blindness, night sweats, bruising easily, craving sweets, acne prior to her period, right index finger and thumb pain, cold feet, sweating easily, swelling of left ankle, bloating after eating, pigmentation on her cheek, frontal headaches, TMJ, and neck and back pain. She was on BCP for two years. She often had clots and menstrual pain before and during the first two days of her periods. Her tongue showed normal color with a red tip and thin white coating. Her pulse was slightly thin, especially the spleen pulse. The differentiated diagnosis of Chinese Medicine was spleen Qi deficiency and phlegm accumulating in three burners.

Cindy's first miscarriage was at eight weeks of pregnancy. She was treated with acupuncture for three cycles. In the middle of the forth cycle, she began trying to get pregnant again. During the treatments, she often felt cold and more comfortable when the room temperature was around 78 degrees. After three months of acupuncture treatments, her period had no clots, and her skin became smoother with less breakouts. However, her temperature did not surge after ovulation. She had severe breast soreness just before ovulation.

Following another two months of acupuncture treatments, her temperature peaked on day 18th and stayed high with clear vaginal discharges on the 17th day. During this cycle, she had a sore throat that came and went and much more phlegm after ovulation with constipation and gas. On the 35th day, her pregnancy test was positive. But her pregnancy hormone level did not go up as was expected.

Unfortunately, she had severe cramps which resulted in a miscarriage again on day 49. After this miscarriage, her nasal congestion and sore throat went away, indicating that her symptoms may have been associated with the hormone changes. Later, the ultrasound showed that the embryo stopped growing at the 5th week of pregnancy. Many fertility specialists suggest that most early miscarriages are associated with poor egg or sperm quality.

Since this was her second miscarriage, she asked to do something else to prevent the next miscarriage other than acupuncture. I suggested Chinese herbs. She started the gentle herbs to strengthen her spleen and kidney function and to transform the phlegm. Later, I found out that she had two cats running around in her house. I tried to persuade her to move away the cats for a while. She told me that she had cats since she had the first baby and that she already developed antibodies against the parasite. After the second miscarriage, her hormone test showed normal level, and the ultrasound did not find any fibroids, endometriosis or tubal blockage. Therefore, it is very likely that these two miscarriages were due to egg or sperm problems. Mother Nature gives our bodies the ability to select the best embryo.

Five months after the second miscarriage, she got pregnant again. Her progesterone level was good. She experienced slight implantation spotting. But at the 6-week ultrasound, the fetus was only at 4 1/2 weeks size, and her pregnancy hormone was more than two thousand. She stopped the herbs because the taste bothered her. She felt tired all the time with chills, increased phlegm, and occasional nausea. These symptoms pointed to the possibility of a good pregnancy. At 8 weeks of pregnancy, the ultrasound did not find a heart-beat; Cindy and her husband were shocked. We did not see each other for three weeks until she had a D&C surgery to move the fetus. I kept thinking about her two cats, but I did not say anything. I know the whole family had a deep feeling about the two cats, especially her lovely daughter. Two months after her third miscarriage, I spoke to a nutritionist about the cats and the miscarriages. This nutritionist told me how the cat parasites can influence the fetus's brain development leading to chromosome damage. I gently advised her to move the two cats to the basement. I analyzed her third miscarriage and told her that her hormone level was good during the third miscarriage, but it seemed that the fetus's development was influenced by some external factors instead of poor egg or sperm quality, since the miscarriage was not spontaneous as the other two miscarriages. One of the factors may have been associated with the two cats. She agreed with my opinion and instantly moved two cats to the basement.

Five months after the third miscarriage, Cindy got pregnant a fourth time. This time she felt slightly different; she started morning sickness on day 35, with severe breast tenderness. Her vaginal ultrasound at 5 weeks showed that the fetus was normal, corresponding to her last period. This time she had nausea and was tired, but she was very happy. Surprisingly, her craving for chocolate was gone. She did not have any acne, although she still had some phlegm with a scratchy and dry throat and postnasal drip with increased progesterone level. At the eighth week of her pregnancy, the fetus's heartbeat was 160 beats per minute. She had nausea all the time, but her skin looked very clear with no acne. Furthermore, her index finger pain was completely gone, and her neck tightness was generally reduced with increased progesterone and estriol, once her progesterone naturally increased. I stopped treating her at the 12th week of her pregnancy.

Two months before her due date, she could not eat. She vomited two to three times a day with a lot of phlegm in her throat, and she could not lie down during the night. Both of her legs and ankles were swollen. Her blood pressure went up slightly. She asked to resume acupuncture treatment. I did one treatment, and gave her some herbs to sip throughout the day to help with her vomiting condition. Finally, at the end of May, she delivered a healthy baby girl.

Two weeks after her delivery, she came back to see me for her hormone rebalancing. With weekly acupuncture, she was able to fall right back to sleep after taking care of her newborn baby in the middle of the night. She did not have any joint pain, but she did gain some weight. I used a group of acupuncture points to enhance her thyroid and ovarian function. Then I asked her if she would like me to treat her pigmentation on both sides of her cheeks. She had 10 treatments and told me if she had known acupuncture could have helped reduce the pigmentation she would have tried it right after the first pregnancy. After the first 12 treatments, we reduced the treatments once a month. Now Cindy does not wear makeup. At 42 years of age, she only needs to put on light powder and her face looks great without wrinkles.

Case Three: Spicy Food and Miscarriage

Marcy was very fit, exercised regularly and ate a lot of fruits and vegetables, but there were two kinds of foods that she loved: spicy food and ice cream. While getting her Master's degree, Marcy would often have a small bowl of ice cream after her spicy lunch or dinner. She had so much energy with the stimulation of her spicy food that she finished her master's degree in two years and met a very nice university researcher. They soon got married and, after a few months, Marcy got pregnant. She normally had a regular period every 26 to 28 days, which could last 8 to 10 days with some cramps, sometimes with infections, like a UTI. She would take an antibiotic for a week, and then the infection would go away. Normally, she started spotting when her period was due. This time, she did not get the UTI, but she never got a full flow of blood, and the spotting kept going for more than 10 days. She decided to visit her gynecologist, and, surprisingly, she was told that she was pregnant.

Marcy had a normal HCG when she was 7 weeks pregnant, but her spotting continued intermittently. She was not worried initially, but the spotting continued through 7 weeks of the pregnancy after she saw the tiny heart beating rapidly. The newly-wed couple was very excited about this little new life, but the spotting worried them. The doctor recommended endometrial biopsy if the spotting continued after 12 weeks. Usually after 12 weeks, when the placenta is well formed and progesterone reaches a very high level, there is less chance of having a miscarriage. Marcy followed the doctor's direction and had an endometrial biopsy when she was 12 weeks pregnant. The endocrinologist could not find anything wrong in the uterine lining. However, after the endometrial biopsy, the spotting became progressively worse, with more and more fresh blood, until she had a miscarriage at 16 weeks.

The sudden drop of progesterone after the miscarriage gave Marcy depression. Every time she saw a pregnant woman passing by or heard that a co-worker delivered a healthy baby, she started crying. After having hope for a new life for 16 weeks, she ended up needing to have a D&C to clean up her uterine lining so that the spotting would stop after the miscarriage. She came to Boston Chinese acupuncture clinic because, after reading the acupuncture and hormone balance book, she thought that her miscarriage might be related to her imbalanced hormones, due to the stress before and after the wedding day.

After her acupuncturist went through her diet and life style, she recommended that Marcy cut down on her spicy food because it can cause inflammation in her uterine lining and produce internal heat. In Chinese medicine, the internal heat can cause a miscarriage, which is why the heat-clearing herb Huang Qin is often used to treat miscarriage. Marcy was told that the spotting could be related to inflammation of her uterine blood vessels. Also, the ice cream with bad fat and sugar can make the inflammation worse. She was so traumatized by the miscarriage that she was determined to change her life style.

It was not difficult for her to stop eating ice cream because it was not part of her diet when she was growing up. By 18 years of age, a person's eating habits are well formed, and it is so difficult to change a childhood habit. Her acupuncturist told her to think about how her blood vessels and nerves become inflamed with each spoonful of the ice cream because the bad fat and sugar can gradually block the small blood vessels in her uterus, her heart and brain, leading to endometriosis, fibroids, a heart attack and dementia. As an intelligent person, she did not want to lose her memory. However, to stop eating spicy food was very difficult. Her acupuncturist told her that she should combine acupuncture treatment while she was trying to cut down on spicy food because the acupuncture can help her produce more endorphins, dopamine and serotonin, which would reduce her cravings for the spicy foods. With each acupuncture treatment, it became easier to reduce the amount of the spicy food she and her husband consumed.

She noticed that her spotting on the days before and after her period became shorter and shorter, and the cramps went away. Furthermore, her skin did not look as red as before; she had been told that she might have rosacea when she visited a dermatologist. Also, a patch of eczema that she had had on her elbow area almost totally disappeared. After 3 months of acupuncture treatments twice a week, Marcy now had a much shorter period, lasting only 4 to 5 days, and was no longer depressed. She decided to have another month of acupuncture treatments before trying to conceive naturally.

Chapter Eight

Polycystic Ovarian Syndrome

What are ovarian cysts?

Ovarian cysts are common among women during their reproductive years. These are fluid-filled sacs, similar to blisters, located on the two almond-sized ovaries on either side of the uterus. Most types of ovarian cysts are harmless and go away without any treatment. Some cysts appear with increased estrogen and disappear when estrogen levels drop during menstruation. When women go through hormone injections for in-vitro fertilization, they tend to develop many small cysts in their ovaries.

Why are ovarian cysts becoming more common?

1.   High insulin level due to high sugar intake: many doctors, including Dr. Steven Horze and Dr. Carolyn Dean,

suggest the connection between obesity coupled with high sugar intake and polycystic ovarian syndrome. The term "Syndrome X" indicates a group of symptoms including hypertension, obesity, high cholesterol, elevated triglycerides, and elevated uric acid, a chemical causing severe joint pain and kidney stones. High sugar intake can trigger insulin secretion and this high insulin level shuts down the body's ability to burn fat, thus moving the sugar into fat cells. Western medicine uses Metformin to regulate blood sugar and menstrual periods for Polycystic Ovarian Syndrome (PCOS). However, there are many factors contributing to PCOS, not just the blood sugar and insulin levels. Metformin can negatively influence the digestive system and cause indigestion.

2.   Birth Control Pills (BCP): many women take birth control pills to treat acne, menstrual cramps, and irregular periods as well as preventing pregnancy. Long term use of birth control pills causes an imbalance between estrogen and progesterone. Women who have been taking BCP for over 5 years usually have reduced menstrual pain, but their cysts never go away; sometimes they can rupture, creating an emergency situation. When they get pregnant, the cysts can grow bigger with the elevated estrogen levels.

3.   Hyperactivity of the sympathetic nervous system: when scientists transected the sympathetic nerve control over ovaries in rats with chemically-induced polycystic ovarian syndrome, rats started ovulating again, and their cysts size was reduced. Electrical acupuncture activates certain muscle-nerve fibers to deactivate the sympathetic nervous system. It also changes the neuronal activities in the spinal cord and induces the release of a number of chemicals important in balancing the sympathetic and parasympathetic nervous over the uterus and ovaries.

What causes hyperactivity of the sympathetic nervous system?

1.   Too much caffeine
2.   Too much stress
3.   Multi-tasking and over-thinking: when the brain has to handle more than two jobs at the same time, adrenal hormones will be released to increase focus, and energy will flow to certain areas of the brain. This explains why women get pregnant more easily when doing relaxing activities.
4.   Too much alcohol or medication compromises liver function. The liver can no longer break down estrogen

and deactivate its effects, so estrogen and its metabolites lead to an imbalance of estrogen and progesterone, preventing the ovaries from releasing healthy eggs.

5. Anti-acid medications which influence the absorption of B-12, causing anemia and malnutrition. Then, the ovaries cannot produce sufficient progesterone to balance the estrogen. Estrogen dominance can contribute to PCOS. Furthermore, the deficiency of B complex can lead to hyperactivity of the sympathetic nerves, contributing to the formation of cysts in the ovaries and other body parts.

6. Over-exercise. Intense exercises not only increase adrenaline levels, but also build too many muscles with higher levels of testosterone. Women need certain amount of estrogen to initiate regular periods and ovulation. Female athletes tend to have very scanty periods.

How can acupuncture help reduce the polycystic ovarian syndrome (PCOS)?

Acupuncture can balance the autonomic nervous system. The nervous system has Yin and Yang to balance each other in order to achieve optimal health. The sympathetic nervous system is in charge of thinking, daily activities, and the fight or flight reaction for surviving. The parasympathetic nervous system is in charge of the body's basic needs: sleeping, relaxing, food absorption, sex, and body repair. With age, the parasympathetic nervous system becomes weaker, making it more difficult to relax and get a good night's sleep, especially in a high-stress environment. Balancing these two systems is not an act of will, hence the phrase "autonomic" nervous system. It is a self-balancing system, vital for organ function like breathing and heart beats.

When the sympathetic nervous system is hyperactive, the blood vessels on the hands, feet and other internal organs not vital for survival such as the uterus and ovaries, are constricted. This guarantees enough blood flow to the brain, heart, and muscles to cope with stress and dangerous situations. Measuring hand-skin temperature can help determine the sympathetic nervous system activity. In the clinic, it's very interesting that many very active and talented professional women with a busy life and successful career have cold hands and feet, even though they are generally pretty healthy. After a couple of acupuncture treatments, their hands and feet are much warmer.

Chen B in Shanghai Medical University observed the effect of electrical acupuncture on hand temperature, ovulation and ß-endorphin level. Their clinical data showed that after electrical acupuncture in patients who resumed ovulation after treatment, their blood ß-endorphin level either declined from the original higher level or maintained within the normal range. On the other hand, for patients who did not resume ovulation, their blood ß-endorphin level was significantly higher than the normal. Interestingly, when ß-endorphin is high, hand temperature is low and the ovulation rate decreases. This may explain why many professional athletes have irregular periods and ovulation problems due to very high level of ß-endorphin. So the "feel good" hormone ß-endorphin cannot be too high or too low in order to achieve the healthy pregnancy.

Does PCOS only happen to overweight women?

Traditionally PCOS is always linked with metabolic syndrome in which women tend to be overweight with high cholesterol and high blood sugar levels. They tend to grow a lot of facial and leg hair, and periods tend to be irregular and scanty. For the past 15 years, I have been treating many fit women with PCOS. They exercise extensively and are very muscular. They may not have a higher level of estrogen, but their testosterone levels are high and the ratio of LH/FSH is abnormal. They have higher metabolism and never gain weight, but they also have a very strong sympathetic nervous system which reduces the blood flow to ovaries. Therefore, they do not ovulate regularly, and the follicles do not release healthy eggs. Instead, the eggs stop developing and create cysts in the ovaries. For this group of women, even if they do not gain weight, a high sugar intake can still induce a high level of insulin, causing inflammation in the ovaries and uterus. They not only need to reduce their sugar intake, but they also need to practice yoga and meditation to balance their nervous system so they can bring energy and blood flow to their reproductive systems.

Endometrium thickness can also indicate the blood flow in the uterine artery. When blood flow to the uterus and ovaries is optimized, the potential for pregnancy increases. Acupuncture can be one of the most effective methods to improve the blood flow to these two internal organs by choosing the right points with the right manipulation.

Why can acupuncture be used to treat PCOS?

Dr. Elisabet Stener-Victorin of Götenborg University in Germany did an extensive study of the effects of electrical acupuncture on the activity of the sympathetic nervous system. She suggested that vasodilatation in the uterus and ovaries may be caused by stimulation of the sensory nerve fibers which inhibit the sympathetic outflow from the brain and spinal cord.

Since nerve growth factor, a chemical stimulating the growth of the peripheral nerves, can also stimulate cyst growth, Dr. Stener-Victorin used nerve growth factor to measure how electrical acupuncture can influence sympathetic nervous system activity. With hyperactivity of the sympathetic nerves, the level of nerve growth factor increased. When acupuncture was applied, nerve growth factor dropped to a normal level. Therefore, by doing electrical acupuncture, sympathetic activity from the brain and spinal cord can be suppressed, helping relax the uterus and bring more blood flow to the reproductive system, aiding in pregnancy and preventing miscarriage.

When women have polycystic ovarian syndrome, their average testosterone and LH/FSH ratio increase. Electrical acupuncture can reduce these two endocrine indicators of polycystic ovary syndrome. Rebalancing the different hormones helps to restore normal ovulation.

Another study done by Dr. Stener-Victorin et al. evaluated the use of electro-acupuncture for ovulation induction on 24 women with polycystic ovarian syndrome (PCOS) whose main symptoms were irregular periods without ovulation. The percentage of ovulatory cycles in all subjects was shown to improve from 15% (in a total of 3 months before treatment) to 66% up to 3 months after treatment. Responsive patients were noted to have significantly lower body mass index (BMI), waist-to-hip circumference ratio, serum testosterone(T) concentration, serum testosterone/sex hormone-binding globulin ratio, and serum basal insulin levels. The sympathetic activity in normal subjects was not affected by acupuncture, so electrical acupuncture does not influence the normal ovulatory cycles. In patients who do not ovulate regularly, especially in polycystic ovary (PCOS) cases, electrical acupuncture can regulate their ovulation.

Chen D. et al in Guangzhou, China did a clinical study on polycystic ovarian syndrome in 2007. 121 cases of polycystic ovarian syndrome were divided into an acupuncture group of 61 cases and a medication group of 60 cases with a randomized and controlled method. The needle-pricking therapy group was treated by needling the sacral region stimulating points on both sides of the spine and lateral points of Dazhui (GV 14), right underneath the cervical 7th vertebra, and the medication group by oral administration of Domiphen and intramuscular injection of chorionic gonadotropin (HCG).

Levels of hormones and symptoms in the patients were measured before treatment, at the 3rd menstrual cycle, and the sixth cycle after treatment. After treatment for 3 menstrual cycles, the level of testosterone and estradiol were reduced; patients had more regular menstrual cycles with regular ovulation and ultrasound examinations indicated the reduction of ovarian cysts in the two groups. However, at the sixth menstrual cycle after treatment, the conditions of the patients in the medication group were returned to the original levels before treatment, while the conditions in the needle-pricking therapy group still kept at the post-therapeutic level; their menstruation and ovulation were restored to a normal state. Furthermore, the ovulation mucosa and the pregnancy rates were significantly higher in the acupuncture group than those in the medication group. Acupuncture had a good long-term therapeutic effect.

In 1993, Chinese scientist, Gong D observed the effect of different acupuncture manipulation on plasma estradiol (E2), testosterone (T), estradiol/testosterone and cortisol in 78 patients with a "kidney deficiency" diagnosis. The results showed that women's levels of estradiol and testosterone were lowered by both reinforcing manipulation (RFM), strengthening certain organ's function, and reducing manipulation (RDM), dispersing the pathogen from the body. The E2/T ratio was lowered by the reducing manipulation, but not by the reinforcing method. There was significance ($P < 0.05$) in the reducing method group and the reinforcing method group in E2/T ratio reduction in the women's groups, but no significance in the men's groups. The level of cortisol was decreased by reducing but not by reinforcing. The results suggest that different manipulations have different effects on the level of sex hormones and cortisol in women with kidney deficiency, but not in men. This research also explains why acupuncture can help reduce the stress when women go through infertility treatments.

What are some treatments for PCOS and self-healing techniques?

1. Herbs for PCOS: in Chinese Medicine, cysts are associated with a phlegm condition and blood stasis. If acupuncture only is used to treat polycystic ovary syndrome, it may take a long time to restore the hormone balance. The following herbs are commonly used to treat this condition according to differentiated diagnostic patterns. A Chinese herbalist will make the right combination according to each individual constitution: Zao Jiao Ci, Chong Wei Zi, Bai Jiang Cao, Hong Teng, Shi Chang Pu, Lu Lu Tong, Xia Ku Cao, Chuan Lian Zi, Niu Xi, Tao Ren, Hong Hua, Hai Piao Xiao, Yuan Zhi, San Leng, Er Zhu, Mu Dan Pi, Zhi Nan Xing, Yi Yi Ren, Shi Chang Pu.

2. When bodies have to deal with high blood sugar constantly, the same amount of insulin cannot efficiently lower the blood sugar to normal level. The body has to produce more insulin, a condition called

glucose intolerance. One of the treatments for glucose intolerance is to use the antioxidant N-acetyl-cysteine (NAC) to restore the body's normal response to insulin. In a study done in Rome, women with impaired glucose tolerance and high blood insulin level were given between 1.8 and 3g of NAC per day; they exhibited statistically significant decreases in total circulating testosterone levels and free androgen index, as well as total cholesterol, plasma triglycerides, low-density lipoproteins, insulin, pancreatic C-peptide, and increased insulin sensitivity.

3. Strengthen the parasympathetic nervous system with electro-acupuncture once or twice a week, 3 months before attempting to conceive.

4. Practicing gentle yoga, meditation or Qigong every night before bed. Simply take deep breaths when lying in bed before getting up or falling asleep when doing acupressure points. This simple technique can be combined with acupuncture treatment to balance the nervous system and bring more blood flow to the ovaries and uterus.

5. Pressing the following points to dilate the blood vessels

and bring energy to the ovaries:

Before ovulation:

CV4: From belly button to the pubic symphysis, divide this area into 5 parts. The CV4 is located on the midline of the lower abdominal area, 3/5 of the way down from the belly button to the pubic symphysis.

CV3: From belly button to pubic symphysis, divide this area into 5 parts, located on the midline of the lower abdominal area, 4/5 of the way down from the belly button to the pubic symphysis.

Kid12: one thumb width on each side of CV3, at the same level of CV3, 4/5 of the way down from the belly button to the pubic symphysis.

After ovulation:

Kid 16: one thumb width on each side of the belly button.

CV7: located on the midline of the lower abdominal area, 1/5 of the way down from the belly button to the pubic symphysis.

CV9: located on the midline of abdominal area, the same distance above the belly button as CV7 is below the belly button.

The following points will help reduce blood stagnation and phlegm condition which are related with ovarian cysts and can be pressed before and after ovulation:

SP10: With the knee bent at the 90 degrees, the point is located on the inner aspect of the thigh, three fingers (index, middle and ring) above the inner edge of the top of the kneecap.

SP9: located on the inner aspect of the shin, in the depression below the knee and the lower border of the tibia.

Kid4: on the inner aspect of the ankle, one thumb width below the indentation between the ankle bone and the Achilles tendon.

6. Stop coffee because it stimulates the sympathetic nervous system and may grow more cysts. Drinking light green tea or white tea can increase alertness and mood instead.

Case One:

Anna was a very warm-hearted 32-year-old woman. She looked very fit and beautiful. But after she moved from the West Coast to the East, she developed severe allergies: nasal congestion, post nasal drip, runny nose and blockage of both ears. All this happened while she was trying to become pregnant. She tried 3 intrauterine inseminations (placing selected sperm into the uterus around ovulation time, called IUI) without success. She had a long list of health problems: eczema, pimples, ear ringing, sinusitis, afternoon headaches behind her eyes, nausea, constipation, abdominal pain, bloating, acid reflux, as well as cold and clammy feet. Recently she noticed her facial hair growing faster than before probably related to all the fertility treatments. Later, it was discovered she did have two small cysts growing in her ovaries. Also, her blood sugar would quickly drop to very low level if she waited too long for a meal.

She tried to get pregnant for 18 months before she started intrauterine insemination (IUI). She had been on BCP for 10 years. She ate lots of sweets and ice cream. The first thing I recommended to her was to avoid all the simple sugars and coffee and try to eat whole grain foods as much as possible. As mentioned, high blood sugar and ovarian cysts are closely related.

The first time I treated her with acupuncture, she felt pressure and a warm sensation in her vaginal area. I knew she may get a pretty good response with acupuncture. She tried one more IUI with acupuncture treatment and felt very different from last 3 IUIs. She did not have any cramps or breast tenderness during the procedure. She noticed that every time she started hormone injections, she spotted most of the time between her periods. Only 3 to 4 days before ovulation and 2 days before her period, there was no spotting even though her menstrual cramps and lower back pain were improved with acupuncture treatment. The symptoms indicated that her ovarian function was abnormal or at least her hormones were not balanced.

After she failed the 7th IUI, her insurance company finally allowed her to start the IVF. For the first IVF, she produced 5 follicles. During hormone stimulation, she had spotting, gas, bloating, indigestion, blockage of the left ear and constipation most of the time. In Chinese Medicine, if maternal digestion is not good, the fetus cannot get enough nutrition to grow in the uterus. With all the digestive problems, of course, the first IVF was not successful. While she was having hormone injections, her ovarian cyst burst causing severe abdominal pain; fortunately, her pain resolved after acupuncture, and she did not end up in the emergency room.

One month after her first IVF, she started the injections again. She got pregnant, but her uterus lining was not thick enough to support the pregnancy. Considering she took BCP for 10 years, her uterine lining and pituitary glands were not functioning normally. BCPs are designed to stop ovulation or create an environment which prevents pregnancy. If a woman takes BCP for more than two years, her hormones are totally changed. If this change also leads to gene expression changes, it will take a long time to reverse the changes.

This young lady was desperate to get pregnant, so she tried three back-to-back IVF procedures between May to October, without letting her body recover and rebalance. During the third IVF procedure, she did not produce even two good follicles. Probably, her ovaries were over-suppressed by BCP. I advised her to cut down on ice cream and other sweet stuff while adding more vegetables. As she did this, her bowel movements incrementally became regular. Although she was very skeptical about taking Chinese herbs, she decided to try them since her reproductive endocrinologist said that her ovaries looked like the ovaries of a 50-year-old woman after her third IVF attempt. Since she was only 33, I was hopeful her ovarian function would improve with herbs and a proper diet. Her husband also tried to cook as often he could to provide enough fiber and vitamins to his wife.

During the three month break from these medical procedures, with the help of acupuncture and herbal treatment, her vaginal discharge became clear and stringy for two days just before her ovulation. Her allergies and constipation were much better. She had no back pain or menstrual cramps. She also had two very late periods with very heavy bleeding. Her fertility specialist said she may have gotten pregnant in a very early stage of one of those two late periods. She was ready for the fourth IVF. This time her new fertility specialist suggested that she try a different protocol.

During the fourth IVF, the BCP made her have heartburn, belching, chest pain and a low blood sugar level with extreme fatigue. Again she continued spotting through the whole procedure. This time, seven eggs were produced. Her foot sweating and nasal congestion had become worse with BCP. On the fifth day after transfer, she bled on and off, felt very warm internally and some fluids accumulated in her left ear with a runny nose. The fourth IVF was not successful.

Now with so many failed IVFs, I convinced her that the raw herbs would be the only way to help her to restore her normal ovarian function. She took raw herbs for three months, her ovarian cysts disappeared. She had regular bowel movements. Her digestion was improving. Her PMS symptoms disappeared: the swollen legs, mood swings, and heavy sensations on her eyelids. She only spotted two days before her period. For the sixth IVF, she started with an estrogen patch instead of BCP. This time she produced 13 follicles, 9 fertilized, two were transferred into her uterus. After the transfer, she had only slight bleeding for half a day. Her pregnancy hormone HCG went up from 458 to more than one thousand 18 days after the transfer. When she told me how tired and cold she felt 10 days after the

IVF, I knew her pregnancy would be a good one. After she got pregnant, she did not want sweets anymore. At the seventh week of pregnancy, her baby's heartbeat was 120 beats per minutes. Her constipation came back with a high level of progesterone, but her allergies were much better with her healthy diet. We stopped acupuncture at the tenth week. A year later after she delivered her first baby, she told me that she got pregnant naturally with her healthy diet and mild exercise.

Chapter Nine

Premature Ovarian Failure

What is premature ovarian failure?

Premature Ovarian Failure, also called Primary Ovarian Insufficiency, is when ovaries lose their function before a woman has reached her normal menopausal age. This disease affects 4% of American women.

Symptoms:

There is no period, or the period comes only every few months, with hot flashes, night sweats, low energy, low sex drive, early onset of osteoporosis, weight gain, vaginal dryness, mood swings, bladder control problems, drooping muscles, arthritis, and sleep problems.

What could cause premature ovarian failure?

1. Viral infection can destroy the pancreatic, ovarian and other glandular functions.

2. Autoimmune disease, when the body produces antibodies to attack the ovaries because the immune function is out of balance.

3. Too many stimulants such as coffee can bring all the energy to the brain and muscles, while the ovaries have insufficient blood and energy flow most of the time.

4. Surgical removal of ovaries.

5. Using birth control pill too early and for too long can make ovaries stop functioning early because "if you don't use it, you lose it."

6. Genetic defects, when certain chromosomes are abnormal.

7. Enzyme function problems or Celiac disease, causing insufficient absorption of vitamins and minerals which influence the production of estrogen and progesterone.

8. Chemo or radiation therapy.

9. Abnormal function of the pituitary or hypothalamus: when these two glands cannot produce the right amount of the hormones to stimulate the ovaries.

How acupuncture can help with this condition?

One theory is that electro-acupuncture promotes the function of the hypothalamus-pituitary-adrenal axis (HPAA) by increasing the synthesis and secretion of one of the adrenal steroid hormones, androgen. This androgen then is transformed into estrogen in the fat tissues and thereby restores the normal function of the HPAA, benefiting the function of the corpus luteum, the residual part of the follicles after the eggs have been released. This is very important in restoring the health of the women with premature ovarian failure, or an insufficiency of the luteal phase after ovulation.

If a woman is diagnosed with premature ovarian failure, her FSH level can be above 30. Theoretically, she has already started "menopause". Can the ovarian function be restored with acupuncture and herbs? It depends on how serious the ovarian function has deteriorated.

In most situations, an endocrinologist would suggest that a woman with premature ovarian failure seek out a donor's egg to get pregnant. With a healthy donor's egg, a woman still needs a healthy uterus, which can provide the best environment for the embryo to grow. In this situation, acupuncture can help improve the blood flow to the uterus, thyroid gland, adrenal gland and ovaries to coordinate all the glands' functions because a healthy pregnancy relies on the functions of all of those glands.

Combining acupuncture and DHEA, ovarian function can be improved:

Drs. D.H. Barad and N. Gleicher of Albert Einstein College of Medicine published a case of a 42-year-old woman whose ovarian function was dramatically improved by using the dietary supplement dehydroepiandrosterone (DHEA) and acupuncture. In a private IVF center, this patient initially presented with severely decreased ovarian reserve. The doctor used medication to induce ovulation while she was taking DHEA and undergoing acupuncture treatment. In her first treatment cycle, her peak estradiol was 1,211 pmol/mL. After seven months of DHEA supplementation and acupuncture treatment once or twice a week, her peak estradiol in cycle 8 was greater than 18,000 pmol/mL. Because of fear of hyper-stimulation, her endocrinologist reduced her gonadotropin stimulation by 25%. In the ninth cycle, her peak estradiol was 9,178 pmol/mL, and she produced 17 eggs that gave rise to 16 embryos. During the 11 months of treatments, the patient had undergone nine treatment cycles while continuously and dramatically improving her ovarian response and banking 66 embryos overall. This case illustrates the possibility that ovarian function may be improved with proper treatments, even in women of advanced reproductive age.

Case one:

Glenda was a 35-year-old woman. Her FSH was 60 and was diagnosed with premature ovarian failure. Her husband had almost no sperm. She was told that she had no hope to get pregnant even with a donor's egg at the infertility clinic. Her friend referred her to an endocrinologist in New York. The specialist in New York got a few sperm from her husband's testicles surgically, and then fertilized her eggs in a test tube. She started acupuncture three months before she went to New York and continued with acupuncture twice a week during the whole procedure. Miraculously, she got pregnant with one of those two sperms. Considering that a normal sperm quantity should be at least 20 million in order to fertilize eggs, this story should give many women hope and confidence to try to improve their ovarian function with acupuncture before starting IVF procedures.

Case Two: Genetic factor only has 30 percent effect:

Gina was a 34-year-old teacher with a passion for children. She had started trying to get pregnant 5 years ago when she was 29 years old. She tried 8 IUI and 5 IVFs without success. Then she was told that she may have premature ovarian failure due to a genetic factor. Her mother had stopped having periods at age 41, which increased the odds that this was true. She tried acupuncture for a year during the 5 IVFs but without luck. When she came to see me, she was ready to accept donor's eggs, which would have been extremely expensive and difficult.

I examined her tongue and pulse and found out that her tongue color and sublingual vein looked good, without blood stasis. She was a little bit overweight. She is a very happy and nice woman. Since she was still young, I advised her to try acupuncture and herbs to balance her hormones while she looked for donors. I also told her to take herbs to prolong her ovarian function so that she can postpone her menopause age.

She was already showing symptoms of peri-menopause: increased FSH levels, waking at 4:30 and unable to fall back asleep, craving sweets, weight gain, headaches, neck, back, wrist and hip pain, and anxiety. She was a healthy eater with a lot of vegetables and fruit in her diet. But she had been on BCP for 6 years to control her heavy period. After so many IVF cycles, her menstrual cycle had recently become shorter, down to every 24 or 23 days. She had had acupuncture for 3 years once a week, yet did not get pregnant, although it did improve her period.

She loved sweets and ate a lot of chocolate. By the time she came to see me, she ate dark chocolate twice a week. Her estrogen level tended to be low, and her husband's sperm was abnormal with less than 40% of the sperms normal.

Her temperature chart showed shortened luteal phase: the basal temperature rose slowly, and then dropped too early. She tended to feel warm, indicating kidney yin deficiency.

She started the Chinese herbs, hoping she could get ready for her donor's egg. In her next cycle, ovulation was on day 17, with clear discharge on day 16; her follicular phase had become longer, allowing her follicles more time to mature.

Her sleep had improved dramatically. She tried so hard to cut down on her sweets, and she got her period on day 26 the first cycle after her acupuncture and herbal treatment combination. But the luteal phase was still slightly short. After her one month of herbal intake, her period became lighter and lasted shorter time, although she still had menstrual cramps. After another month of taking the herbs, she noticed that she became very hungry in the morning, and she tended to get lightheaded during morning exercise. I suggested she practice yoga and Qi Gong and go for walks instead of intense exercise.

She took two formulas. Formula 1 was before her ovulation to invigorate the circulation in the ovaries and uterus. Formula 2 was for after ovulation to bring up her progesterone and relax her uterus to get ready for implantation. With Formula 2, and her basal temperature jumped from 96.6 to 97.8 to 98.2. She became happier and more energized.

After her two months of herbal intake, her period came on day 28. She had a two-day heavy flow with cramps and a lot of clots, and the bleeding stopped for one day, with another day of full flow. She was very tired with facial swelling even though her sleep has been good.

The following cycle came on day 18, with a very short period, mood swings and unhappiness before the period after she ate a lot of sweets. This earlier and shorter period indicated that her diet played a very important role in her already compromised ovarian function. Sweets can harm ovarian function because it causes inflammation in the blood vessels and ovaries. She was determined to stop sweets.

After the very abnormal cycle, she did not get her period on the 29th day, but she had mood swings and headaches, indicating that her estrogen and progesterone had gone up. On day 35, her HCG was 1932. Soon after, she had a vaginal ultrasound, which showed two sacs in her uterus. She was tired and felt cold all the time, which were all good signs for a good pregnancy. She was so thrilled and could not believe she could get pregnant by herself by doing acupuncture and herbs for only four months. She even called me and asked if the herbs could cause false positive result in pregnancy test. The answer is no.

On the eighth week of her pregnancy, ultrasound showed two normal heartbeats. She had morning sickness, and a little spotting, which scared her so much that she decided to continue her acupuncture until the babies are delivered. With two fetuses growing inside her, the dramatically increased hormone levels made her very sensitive to smell.

Interestingly, she did not crave for sweets anymore but instead craved salty foods. She had loose stools three times a day instead of her constipation in her first trimester, which is quite different from the majority of the pregnant woman. By week 13, Gina had no nausea, and ultrasound showed a boy and a girl. She continued her acupuncture treatment during the whole pregnancy. She only had slight back and hypochondriac pain. She ate a very healthy diet during the whole pregnancy. By the end of 35 weeks, her legs were swelling, but after a few acupuncture treatments, the swelling went down dramatically. At the 36th week, ultrasound showed boy was breeched, so we added some points to make the baby move. After four weeks, both babies were in the right position.

Of interest, when her progesterone went up, her sleep pattern also changed; she was able to sleep from 10pm to 6am instead from 9pm to 4am. At the end of the pregnancy, she started having hand pain, swollen ankles and heartburn due to the water retention and changed digestive function with increased progesterone levels. She began acupuncture every week again to reduce her hand pain and acid reflux. Just before her due date, her sleep pattern changed again: she would sleep 4 to 5 hours, wake up for two hours, and fall back to sleep again. She continued with her acupuncture treatment to the end of the pregnancy. Gina delivered two healthy babies.

Two months after she delivered her babies, she started acupuncture treatment to reduce the carpel tunnel syndrome resulting from pregnancy. She was worried she would lose her period after her two babies were born, so she came back to take herbs and acupuncture to restore her progesterone levels. With two newborn babies, she could only come every two weeks for acupuncture treatment when she had a babysitter. Her period now came every 24 to 26 days, without hot flushes or night sweats. She noticed that if she had fewer sweets and more vegetables, her luteal phase could reach 14 days. During Christmas time, her period was shortened to 22 days with more cramps and breast tenderness due to the diet changes.

Now Gina is 44 years old, and she still has regular periods with 6 to 7 hours of sleep. With regular acupuncture treatments, her wrist pain is almost gone. She is a truly believer of Chinese Medicine and acupuncture treatment. One time her twins had a cold when they were toddlers, so she gave them some herbs, and the coughing and sore throat went away within two days. Now she usually has two bottles of herbs: one is for cold and flu, and the other is for her hormone balance.

What can be done to restore or improve ovarian function?

1. Consuming sufficient raw vegetables and fruit to provide the necessary nutrients for restoring the

ovarian function. Even though the ovaries do not produce sufficient estrogen and progesterone, the adrenal glands can still produce some estrogen, testosterone and progesterone. The green juice made from kale, lemon, apple, spinach and berries will help adrenal gland produce more estrogen and progesterone, which can help improve ovarian function.

2. Reducing stress with meditation, Qi Gong, Tai Chi and yoga will help the adrenal gland and ovarian function. Even menopausal women can release eggs when all her hormones are in good balance, which is why some 50 year olds can get pregnant occasionally when they are in a stress-free environment.

3. Taking Chinese Herbs can bring the energy to ovaries and digestive system with some herbs such Yin Yang Huo, Xian Mao, Shan Yao, Goji berries have phyto-estrogen effect, creating an environment beneficial for egg production.

4. Getting electrical acupuncture twice a week for two to three months to improve ovarian, pituitary and hypothalamus functions. Once these three glands

are in good condition, there may be ovulation every two to three months. Progesterone injections can induce a period initially, but it never can improve ovarian function. After few years, the injections will stop working.

5. Regularly eating fermented soy such as tofu will also help restore the ovarian function because tofu has phyto-estrogens which benefit the ovaries.

6. Stop eating sweets such as ice cream, chocolate and cookies to reduce inflammation in the ovaries.

7. Cooking this porridge of herbs two to three times a week: Yi Mi 40g, Shan Yao 30g, Go Ji 15g, Gui Yuan 12g, adding sufficient water.

8. Drinking Raspberry tea every day, or add wheat grass in your green drink.

9. Massaging the following points to bring the energy to uterus, kidneys and ovaries: Using the palm of your hand as a measurement unit, from the belly button, measure one palm down. This point stimulates the reproductive organs, uterus and energy production. From this first area below the belly button, measure right and left one palm. There

are two points in these areas that stimulate the ovaries, so the massage area is about the size of your palm clockwise. The area around the belly button is important to stimulate energy for the reproductive system as well. Massage around the belly button moving clockwise. Lastly, on the inside of your ankles, in the divot between the ankle bone and the Achilles tendon, there is an important point to stimulate energy and reproductive organs. You can massage each of these points for one minute 2-3 times daily.

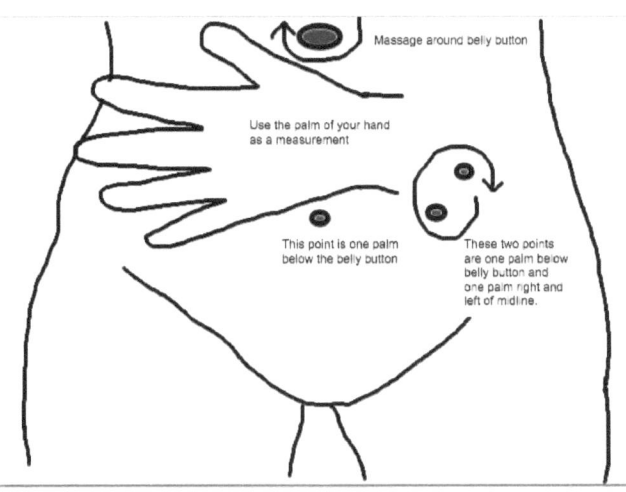

Chapter Ten

Unknown Infertility

Infertility from an unidentified cause accounts for 15% of all infertility cases. Couples with this situation can be especially frustrated, particularly because they never know if their situation is changeable or permanent. It is usually termed "Infertility of Unknown Causes", "Unexplained Infertility", or "Unknown Infertility". This does not mean that there is no reason for this category of infertility, but rather it cannot be diagnosed. The percentage of couples falling within this category will decline as new causes of infertility are discovered.

The following are several situations frequently seen in clients with unknown infertility:

1. Low blood pressure: when one's blood pressure goes too low, the body will shut down the blood flow to the uterus or ovaries in order to provide enough blood to the brain, heart and other vital

organs. The following may help regulate the blood pressure: drinking green tea in the morning two hours after breakfast, avoiding simple sugars, eating every 3 to 4 hours, taking a nap after working more than 5 hours to rejuvenate adrenal gland function to avoid the drop in blood pressure, drinking enough water during the day, and sleeping well, which is very important to keep the adrenal glands and ovaries functioning normally.

2. Thin uterine lining due to poor circulation: after a large amount of blood loss, ovarian function can be compromised. If the ovaries do not produce enough estrogen and progesterone to stimulate the growth of the uterine lining, the fetus cannot grow properly in the uterus.

3. Indigestion and anemia: if anti-acid medications are taken for a long time, it can cause a deficiency of Vitamin B12 and malabsorption of essential amino acids, leading to anemia and poor circulation to the uterus. Sufficient vitamins and minerals are also vital for the production of progesterone.

4. Hypothalamus-pituitary-ovarian dysfunction: if a woman has been taking birth control pills (BCP) for

more than five years, she might not ovulate normally, and her uterus lining could become too thick for pregnancy. She could have problems getting pregnant or miscarry after she stops BCP. Moreover, there may be miscommunication between the hypothalamus, pituitary and ovaries caused by BCP. Then why do some women get pregnant right after they stop BCP, while others have difficulty? The answer depends on one's lifestyle and genetic makeup. If one eats very unhealthy food and never exercises, stress levels are too high, or one has other sub-clinical problems, then more time is needed to restore the balance of the reproductive system.

5. The adrenal glands also can produce small amounts of estrogen and progesterone. If the adrenal glands are depleted, the ability to conceive can be affected. Also, blood pressure and blood sugar will fluctuate a lot, and one's ability to adjust to environmental changes will diminish. When I had my first child, I had blood loss for a very short period of time. I almost fainted at that time. My adrenal gland function was compromised afterwards. At the same time, I was studying for my PhD for 6 years, so I

could not get pregnant for 7 years, even though I had regular periods and was only 35. After I got my Ph.D. and my stress level was lowered, I got pregnant within a year after I moved to Boston.

6. High stress can deplete progesterone levels because progesterone shares the same precursor with cortisol. If one's uterus, ovaries and fallopian tubes are all normal, and hormones are within a normal range, high stress and anxiety can switch the vital energy away from the uterus and ovaries. The body is in a "fight or flight" condition to survive, and the reproductive system will shut down to save energy and prioritize the response to the stress.

7. If one's body produces antibodies to sperm, then the egg cannot be fertilized or implanted in the uterus. Immune function needs to be in balance to allow the fetus growing in the uterus.

8. If men do not drink enough water, or if they take anti-histamines for prolonged periods, the sperm may not be able to swim fast enough to reach the uterus before the egg dies. The healthy sperm can survive for five days, but the egg can only survive one or two days.

9. The vaginal pH is not optimal for sperm survival. Too much meat or sugar can create acidic condition.

Case One: Difficulty with Second Pregnancy

Hannah is a 39-year-old woman who tried IVF 3 times. She developed many food and environmental allergies, including peanuts and dust, when she was a child. She experienced anaphylactic shock if she smelled a trivial amount of peanuts in her food. She had eczema and asthma when she was a child. Since she ate mostly homemade food during childhood, she grew out of the asthma after puberty. She generally felt cold with sinusitis and digestive problems. She bruised very easily. Her skin was very dry most of the time. She had cold hands and feet and a slight headache before her period. She also had fibrocystic breasts. Her period had been very dark colored with lower back pain.

She did not respond to hormone stimulation very well. During each procedure she only produced three follicles. Furthermore, the hormone stimulation triggered her eczema. Her insurance refused to pay for the fourth IVF, and she was frustrated with overcrowded fertility clinics because the physicians had no time to listen to their patients.

After a couple of acupuncture treatments, her eczema was much better. While she was waiting to change her insurance, we added a Chinese herbal formula to her treatment protocol. According to her symptoms, her diagnosis was blood deficiency and stagnation, kidney and spleen Yang deficiency. Three formulas were used for each cycle. The first one was to tonify and invigorate her blood and warm her cold uterus; the second one was to tonify her kidney Yin and Yang to optimize her ovulation, and the third one was to smooth live qi and strengthen her spleen function to create a hospitable environment for the embryo to grow. Due to her allergy to peanuts, she could not take Tao Ren (peach kernel). With Tao Ren, she had a very sticky and itchy sensation in her throat. I modified her herbal formulas by choosing other blood invigorating herbs.

After two months of herbal treatment, her bloating was dramatically reduced and her eczema was completely gone, but her tongue was still pale and slightly purple. Interestingly, the tongue color changed after her ovulation.

She tended to have acid reflex, so I added Wa Leng Zi to address this issue.

After 3 months of treatments, her physical check-up showed that the thickness of her uterus lining was good, and her FSH was lowered from 11 to below 7. Her body temperature showed a beautiful biphasic chart. However, she still did not conceive. I added more tonifying kidney Yang herbs; she felt nauseous and a crampy sensation, as if she had gotten pregnant, after ovulation. She was so worried about the reaction that she called me.

She soon called me to tell me she was pregnant. I was overjoyed for her. Her husband said that the next step was to find the heart beat. I said that this has to be a healthy pregnancy after all our efforts. There should be no doubt that the heartbeat would be normal. Her pregnancy hormone jumped up to 20,000 within two months of pregnancy, and the fetus's heartbeat was absolutely normal at the sixth week of pregnancy.

She stopped herbs but ate ice cream every evening with her husband during the pregnancy. I thought that this could damage her digestive function. Since she had placenta previa, she had scheduled a C-section. She lost a lot of blood during the C-section and had anemia for a while.

A couple of months after her charming daughter was born, she started Chinese herbs and did not have any hot flashes, night sweats or insomnia, as most women do. Her FSH was even lower than before the first pregnancy. Her physician never rechecked her anemia condition until 3 years after her daughter was born.

I thought that, since her FSH was quite low and she had already gotten pregnant with acupuncture and herbal medicine the first time, it would be easier in her second pregnancy, but this was not the case. She tried nine-months of acupuncture and herbs, and her hormone checkups had all been normal, but she could not conceive. Finally, she decided to do IVF again, hoping the hormone stimulation could create a miracle this time. For this second pregnancy, she had another couple of chances to try IVF before age 42.

Hannah started her first IVF cycle after her daughter turned one. She faithfully took the herbs until one week before her hormone stimulation. Her FSH was 7.5, but her estrogen was slightly high. She started with an estrogen patch, then an estrogen antagonist (the estrogen blocker). Around the area she applied the estrogen patch, she developed big rashes and felt itchy all over her body. She produced 3 follicles; one was very big, and the other two were quite small. The endocrinologist waited for another two days to see if more follicles were developed.

Strangely, her follicles disappeared; there was no follicle to be collected to fertilize the sperm. After all the hormone injections, her IVF was cancelled, and she had an IUI instead. After the cancelled IVF, she started having night sweats and severe menstrual cramping. I added Chao Mai Ya and Gui Ben Jiao to enhance her digestion and rebuild her blood and kidney Jing in her first and second formulas. Her eczema also started after hormone stimulation, and her cycle became shorter and shorter.

A couple of months after her cancelled IVF, she had inflammation of her pancreas with right lower abdominal pain, indigestion and acid reflux. At first, she suspected that the herbs were the cause of these symptoms. I recommended that she stop taking herbs for two months. She also took anti-acid medication in the meantime. Her symptoms still bothered her after she stopped herbs. I suggested that she eat an early dinner and drink more water before each meal. Later, with the change in her eating habits, she was able to stop the anti-acid medication, and her symptoms were completely gone. However, after she had stopped herbs for two months, her period became irregular, varying between 24 to 47 days without a hormone surge and ovulation. She wanted to start herbs again before she started the next IVF.

Since during the previous IVF her three follicles had disappeared, her insurance company refused to pay for this IVF, saying she was a poor responder to hormone stimulation. She had to convince them that she did technically have 3 follicles. Her endocrinologist wrote a letter to support her appeal while she continued her acupuncture and herbal treatments with me. After taking herbs for two months, her shortened luteal phase became normal.

Four months later, the insurance company finally agreed to pay for another IVF cycle. The new IVF started with BCP, which made her breasts tender to the touch. This time she produced 5 healthy follicles with the right thickness of uterine lining and a normal hormone level. Five decent embryos were transferred. Two were rated as super, two were good and one was average. Furthermore, her husband's sperm was now 90 million, improved since having twice-a-month acupuncture for a couple of months.

Hannah had acupuncture before and after her embryo transfer. With this excellent response to hormone stimulation, there was concern about her having triplets. Unfortunately, she did not get pregnant. This puzzled her physician. They could not figure out why she failed the second IVF. Finally, her fertility specialist decided to examine her uterus with certain equipment to determine if she had endometriosis or polyps. They found out that the back of the uterus had a very thin lining at the location where the 5 embryos were transferred. Only a very small portion of her uterus lining close to the uterus cervix looked fairly decent; that was the place where her first pregnancy had occurred. Surprisingly, there was still a piece of placenta stuck inside the uterus around the cervix. She also might have endometriosis. Her physician cleaned up her uterus and got that piece of residual placenta out. She also started taking an estrogen pill for 8 weeks to build up her uterine lining. After the D&C, she continued with her

Chinese herbs and acupuncture for another three months, and she finally got pregnant with a third IVF.

In this case, why were all of her hormone check-ups normal, but her uterus lining too thin? The normal range of the hormones is very broad. Hannah's hormone levels may have been lower than before, but still falling within the normal range. As mentioned, she had a lot of digestive problems and low blood pressure. She also had to commute to work for an hour every day after she had her first child. The above factors would contribute to the chronic anemia and insufficient blood flow to the uterus and ovaries. Even if she took herbs twice a day and had acupuncture once a week, it is not sufficient to offset the influence of daily stress and the late dinner-related indigestion.

Her difficulty conceiving the second time may be associated with the following possibilities: her severe blood loss during the C-section may have resulted in insufficient blood flow to her uterus and ovaries. The ovarian function was severely compromised by this sudden blood loss. That is why her progesterone was not sufficient to maintain optimal thickness of her uterus lining so that suddenly her three follicles were gone. Plus, the residual placenta in her uterus could also make the implantation very difficult. When her ovarian function was suppressed during the IVF procedure with the BCP, it just could not recover fast enough to support the pregnancy, considering that her ovarian reserve could drop sharply because of her age. Lastly, the 10 months ice cream during her first pregnancy could have also contributed to the endometriosis.

Self-healing Technique for unknown infertility:

1. Optimize blood pressure by drinking green tea in the morning two hours after breakfast, avoiding simple sugar, and eating every 3 to 4 hours. Taking a nap after work or lunch can rejuvenate the adrenal glands, preventing adrenal insufficiency and stabilizing blood pressure. Drinking enough water and raspberry or cinnamon tea during the day can also help blood pressure. Low blood pressure can protect the heart, but it can also sacrifice ovarian function. Very thin women with extremely busy schedules tend to develop this condition. Taking Chinese herbal formulas to strengthen kidney and spleen function will optimize blood pressure.

2. Prevent high blood pressure by avoiding too much salt which can develop into arthrosclerosis, thereby compromising blood circulation to the ovaries, brain and uterus.

3. Reduce stress with weekly acupuncture treatments, daily yoga, Qi Gong or meditation.

4. Rub the following acupressure points for 30 seconds to a minute when meditating so oxygen and nutrients circulate better to the ovaries and uterus:

   Yin Tang: the midpoint between the eyebrows, above the nose on the medial side of eyebrow.

   GV20: the top of the head and the midpoint on an imaginary line connecting the tops of the ears.

   Heart channel points: with the palm facing up, in line with the pinky, rub on inch from the wrist crease toward the elbow. You can check these points on www.bostonchineseacupuncture.org.

5. When it is not near the time of ovulation, condom use can let the uterus avoid direct contact with the sperm, so that, after a few weeks, the body will stop producing antibodies against the sperm.

6. Avoid foods which trigger histamine release, such as spicy foods, alcohol, and chocolate. Getting rid of inflammation in the body can facilitate a healthy pregnancy. Too much histamine can also lower blood pressure and reduce the blood flow to the uterus and ovaries.

7. Taking Chinese herbs such as He Shou Wu, E Jiao, Shu Di Huang and Gu Ji Zi to build the uterine lining. Find an herbalist who specializes in treating infertility to write a formula to improve the ovarian, uterine, adrenal and thyroid functions.

Chapter Eleven

Acupuncture and IVF

IVF, or *in vitro fertilization*, has existed for more than 40 years. The process involves stimulating a woman's ovaries to produce multiple follicles which are then harvested from the ovary and fertilized outside the womb with sperm. After growing for several days, the embryos or sometimes blastocysts (older embryos) are transferred to the uterus. After the transfer, implantation should occur between 4 to 8 days.

Before the implantation happens, the fertilized egg will move around the uterus and fallopian tubes. Scientists have been trying to invent a kind of glue to fix the embryo in the uterus so it will not move, but this idea has not been successful. There is even a procedure called endometrial scratching which can help the lining be more receptive to implantation. If a woman does not create a healthy and hospitable uterine environment or if she has a lot of adhesions due to the pelvic infection, fallopian tube blockage, has had a D&C or tubal ligation, then miscarriage or ectopic pregnancy can occur at higher rates than natural pregnancy.

IVF is a very stressful procedure. Women have many hormones injected into their body during IVF. The ovaries are stimulated to develop more than 10 follicles in one cycle, and each follicle releases large amounts of estrogen which can influence the immune function and nervous system. Estrogen can lead to hyperactive immune and nervous systems, causing the immune system to produce auto-antibodies which then attack the fetus. The hyperactive nervous system can release stress hormones such as adrenaline and histamine, leading to anxiety, allergies, and insomnia.

These hormone changes can profoundly influence a women's health even after delivering the baby. Furthermore, the hormone imbalance can lead to a miscarriage and abnormal development of the baby's brain. For instance, lower levels of progesterone may be linked with autism. As stress levels grow, the progesterone levels go down. Lower levels of the progesterone can cause women to develop autoimmune diseases such as lupus, colitis, Hashimoto's thyroiditis, Crohn's disease, and rheumatoid arthritis.

Research led by Dr. Wolfgang in Germany revealed that young women typically have a 20 to 30 percent chance of conceiving during IVF, but the pregnancy rate falls to less than 10 percent for 39-year-old women, and then to only three percent for women at age 44. But acupuncture can double the success rate by relaxing the uterus, improving ovarian function and balancing the hormones so the embryo can implant in a very hospitable environment. Other research done by a physician in Minnesota indicates that acupuncture can reduce the miscarriage rate and increase the rate of successful pregnancies during IVF cycles. These success rates have influenced common practice in IVF. For example, every infertility center in Boston has its own acupuncturist to perform acupuncture before and after the embryo transfers.

How can acupuncture benefit IVF procedures?

Reduce the stress level: when the sympathetic nervous system is hyperactive, the blood flow of the ovaries and uterus decrease, which may compromise the ovarian function. Acupuncture can reduce the sympathetic tone.

Having acupuncture once or twice or a week before IVF can help to reduce the side effects of hormone injections. Most hormones stimulate the body to produce a lot of heat, which can tip the balance of many body systems. Acupuncture helps to clear up this internal heat and balances the immune, digestive and endocrine systems.

Before or after transfer, acupuncture helps relax the uterus to facilitate the implantation and prevent miscarriages.

During the period, acupuncture can help the body clean out the uterine lining so the old blood can be discharged thoroughly, which then helps the uterus grow a healthy lining.

Acupuncture can help the adrenal glands and ovaries produce more progesterone. This is very critical for early pregnancy. Without sufficient progesterone, there is a higher chance for a miscarriage. Synthetic progesterone medication is not the same as the natural progesterone.

Acupuncture can help balance the immune function to reduce the production of the auto-antibodies.

What can be done to increase IVF success rates?

Three months before IVF:

1. Stop drinking coffee or switch to green tea or decaffeinated coffee because only one cup of coffee can increase the miscarriage rate, according to an NIH study. Caffeine can also switch on energy to certain brain areas, leaving the ovaries and uterus without sufficient blood flow. It can also stimulate insulin release so the blood sugar is transferred to fat cells instead of being used for energy and maintaining the body temperature. When women drink coffee regularly, they not only have more anxiety but also crave more sweets, and the high insulin levels can cause inflammation in their ovaries and uterus as well as the blood vessels and nervous system.

2. Eat more cruciferous vegetables such as cabbage, cauliflower, kale, and Brussels sprouts. The best practice is to use a high power blender to make homemade green juice. The vitamins and minerals in the green juice help reduce inflammation in the

ovaries and uterus. Also, the cruciferous vegetables can serve as a natural aromatase blocker to prevent breast, uterine, ovary and even prostate cancer; aromatase inhibitors block high levels of estradiol (the bad estrogen) production which normally stimulates uterine fibroids, endometriosis, or an enlarged prostate.

3. Stop drinking wine or any kind of alcohol because alcohol can enhance the aromatase function, causing an estrogen dominance condition. Alcohol also can produce a lot of heat, leading to inflammation and even miscarriage. Alcohol can also cause inflammation of the digestive system so the stomach cannot absorb vitamins and minerals efficiently. Also, it compromises liver function, the body's cleaning system, so the "bad" estrogen will stay in the body for a longer time.

4. Start using wheat grass supplements, 3 tablets in the morning with 8oz water; then in the afternoon, drink raspberry leaf tea. The wheat grass and raspberry tea can help balance hormone levels and reduce the stress hormones. During IVF cycles, it is easier to develop the estrogen dominance condition.

With these two food supplements, one can produce healthy eggs and also have less chance to gain weight. When estrogen and progesterone are in balance, the embryo can implant and grow healthily. I have treated a few pregnant women who developed breast cancer due to the hormone imbalance and unhealthy diet. It is unfortunate when a pregnant mother develops cancer because the high hormone levels during the pregnancy makes the cancer progress much faster.

5. Taking Chinese herbs to improve digestion and ovarian and uterus function is beneficial. The majority of Chinese herbs have been used to treat infertility for five thousand years and are very safe. Women who have repetitive miscarriages can take Chinese herbs even when they are pregnant. Finding the right herbalist who has experience in treating infertility, fibroids, endometriosis and premature ovarian failure is extremely important. If the husband can take the herbs at the same time, testosterone levels and sperm quality can be improved. A healthy pregnancy achieved with one IVF saves money and future health problems.

6. Starting acupuncture once or twice a week will improve the reproductive function. Also, doing some acupressure points at home is helpful. Acupressure on the foot and abdominal area will strengthen kidney and spleen function and smooth liver qi. Find a local, qualified acupuncturist to teach some acupressure points. The points will be different before and after embryo transfer.

7. Learn some acupressure points from my YouTube channel, Boca Raton Acupuncture Clinic, to reduce the stress and produce more progesterone as well as balancing immune function during the IVF procedure.

8. Add wild yam, daikon and fresh ginger to one's diet to enhance digestion. Fresh ginger serves a natural blood thinner to prevent the blood clotting which can occur with hormone injections.

9. Avoid lobster and spicy foods, as they can trigger histamine release and induce inflammation and allergies. During IVF, spicy foods can cause uterine bleeding and miscarriages, because all the hormone injection can produce a lot of heat and deplete our kidney Yin. Spicy food also can cause hyperactive

immune function, helping produce the antibodies to attack the fetus and the mother's own tissues.

10. Avoid eating any kind of nuts including peanuts, almonds, walnuts and other tree nuts. These are hard to digest and can also trigger immune reactions. Many women who have a vegan diet eat so many nuts that they increase their chances of developing autoimmune problems such as lupus, eczema, Crohns's disease, and rheumatoid arthritis. After delivery, avoiding those nuts will protect the mother from allergies and autoimmune disease.

11. Practice meditation and gentle Yoga three months before IVF to help balance immune function and the nervous system. When one is relaxed, sleep is deeper, and progesterone levels are optimized.

12. Before transfer, take some seaweed to provide enough minerals to optimize thyroid function and reduce ovarian cysts.

13. If the luteal phase is too short, chase berry or L-Arginine can be taken to balance the hormones.

14. Moxa on acupuncture points CV4, CV3 and uterus points can enhance the ovarian and uterine blood

flow before starting IVF procedure. After embryo transfer, moxa can be done on points ST36, SP4 and Kid3.

15. Femoral massage: before the transfer, press the artery in the groin for 30 seconds a couple of times a day to temporarily keep more blood in the lower abdominal area.

16. Eat some pineapple to facilitate the implantation.

17. Avoid sitting in the hot tub or bath after transfer.

18. Avoid swimming in cold water, which might cause uterine contractions, prior to finishing the first trimester.

19. If one has polycystic ovarian syndrome, one should avoid sweets and a high salt diet which can increase cyst growth and facilitate yeast infections; the suppository progesterone can induce yeast infections.

20. Take a two to three month break between each IVF cycle so the body can restore its hormone balance after all the hormone stimulation.

Before, during or after IVF, healthy ovarian and uterine function can be maintained by:

1. Avoiding ice cream and all the other sugary food; high sugar causes high insulin, which promotes polycystic ovarian syndrome. Also, high sugar creates acidic cervical mucus, which may be hostile to sperm.

2. Do not drink fluids while eating. Digestive enzymes are altered by drinking fluids while eating, which can cause indigestion or acid reflux.

3. Avoid extremely salty food because high salts can cause water retention and promote the cyst growth.

4. Do not overeat spicy foods; it will produce too much internal heat, leading to inflammation or autoimmune diseases such as Hashimoto's disease.

5. Do not use tampons. It is not healthy to keep the blood inside the vagina leading to backflow of the blood into the fallopian tubes, one of the causes of endometriosis.

6. Reduce stress by doing yoga, Tai Chi and Qi Gong to redirect energy to the ovaries and uterus. The reproductive system needs to be taken care of in this very stressful time.

7. Try some herbs such as raspberry tea or chase berry tea to balance estrogen and progesterone.

8. Chinese herbal medicine has been proven to treat infertility effectively for more than five thousand years with very minimal side effects. After finding a good Chinese herbalist, take herbs for three to six months before starting an IVF procedure.

9. Acupuncture is a quick way to improve the blood flow to uterus, ovaries and adrenal glands. Try acupuncture at least once a week for 3 to 6 months before starting IVF.

10. Avoid over-exercising. Over-exercise will lead to an adrenaline release which can cut down the blood flow to the uterus. When personal trainers have an extremely fit body, their ratio of fat/muscular tissue

is too low for a normal estrogen level. They usually have irregular or very light periods and infertility.

11. In polycystic ovarian syndrome, it is important to take magnesium, alpha-lipoic acid, chromium and conjugated linoleic acid, which help increase insulin receptor sensitivity within the cell to prevent hyperglycemia, or high sugar levels.

12. Optimize serotonin levels: Serotonin helps to prepare the uterus for pregnancy. If one has general body aches, increasing depression, irritability, premenstrual symptoms, early morning insomnia, night sweats, or panic attacks, eating certain foods can stimulate the body to produce more serotonin. Some of the better-known foods that contain tryptophan, an amino acid used to produce serotonin, are turkey, bananas, milk, yogurt, eggs, meat, nuts, beans, fish, cheddar, gruyere, and Swiss cheese. A better late night snack would be bananas, not ice cream.

13. Optimize one's weight: being overweight can cause too much estrogen to be stored in the fat tissues.

This can tip over the delicate hormone balance, causing the uterine lining to be too thick and inhibit implantation.

14. Women are not marrying until they are in their late twenties or thirties, and women are postponing having their children until even later. After 40 years of age, women do not naturally ovulate every month. With age, there is less ability to compensate for hormone imbalances induced by stress. According to some statistics, a woman's fertility peaks in her early twenties and starts to decline as early as age twenty-seven. By the time a woman is thirty-five, her chance of conceiving decreases by 50 percent, and it shrinks to 20 percent by the time she hits forty. If BCP use continues after age thirty, one may have a difficult time getting pregnant.

15. After age 40, women's hormones fluctuate a lot. In Chinese Medicine, the kidneys and spleen tend to be deficient. Women can supplement their diet with blue-green algae, wheat grass, and green power foods. They should avoid the use of dry, pungent and acrid spices. Those foods can deplete the

kidney essence and make the body produce more internal heat.

16. Environmental pollution: xenoestrogen can create the estrogen dominance situation. The body must be protected from the chemicals as much as possible. Remember, all the chemicals have to be detoxified through the liver and kidneys. Avoiding unnecessary medications can help to preserve the liver and kidneys.

17. Have regular bowel movements and drink enough water to detoxify the body and protect liver and kidney functions. This will also help avoid an estrogen dominance condition.

18. BCPs are designed to stop ovulation or make the uterine lining unsuitable for implantation. Dr. Carolyn Dean indicates that long-term use of BCP is one of the factors causing the rise of infertility rates. The brain gets incorrect information that either the body is already pregnant or the ovaries should not release the eggs, so eventually the

endocrine system cannot communicate in the right way.

19. Get enough sleep: in deep sleep, immune and reproductive systems repair themselves.

20. Avoid multi-tasking: doing two things together triggers an extra adrenaline release which constricts the blood vessels to the uterus and ovaries.

Chapter Twelve

Secondary Infertility

Secondary infertility is defined as inability to conceive or carry a baby to full term after a previous successful pregnancy. More and more women have been diagnosed with secondary infertility over the past 10 to 20 years, even though they had been able to get pregnant previously right after they stopped birth control pills. There are many factors contributing to this frustrating condition.

What are some of the causes of Secondary Infertility?

1. After the first pregnancy, some women may develop endometriosis and fibroids due to hormone imbalances. Progesterone tends to be very low right after delivering a baby. If women do not eat the right foods and drink too much coffee, it is much easier to develop an estrogen-dominance condition, which can

144

lead to endometriosis, fibroids, and PCOS (Polycystic Ovary Syndrome).

2. Repetitive pelvic infection can lead to tubal blockage in which the sperm cannot get to the ovum in time after the ovulation.

3. During the first pregnancy, some women do not eat a healthy diet and develop glucose intolerance, which can lead to PCOS and inflammation of uterus and ovaries. During pregnancy, many women crave sweets such as ice cream or salty foods, causing more ovarian cysts to grow. Each cyst can release testosterone, which interrupts the ovulation process.

4. After delivering the first baby, women tend to develop autoimmune disease or allergies, and the immune system can produce the antibodies to attack the embryo, rooted in a lack of progesterone.

5. If a woman lost a lot of blood during delivery, uterine conditions can be altered, and ovarian function is compromised.

6. Multiple IVF procedures can lead to ovarian dysfunction due to overstimulation of the ovaries.

What can be done to treat secondary infertility?

1. Balance hormones with green juice if there is an estrogen-dominance condition. Delicious smoothies can be made with kale, lemon, apple, coconut water, spinach and fresh ginger, or a simple one with kale and apple. Kale belongs to the cruciferous vegetable family and can block the over-production of bad estrogen: estradiol, to balance the estrogen and progesterone. If the estrogen-dominance condition is corrected, the fibroids, endometriosis condition can be corrected.

2. Drinking Rosemary leaf tea, one or two bags each day, can also help balance estrogen and progesterone. Rosemary tea can also suppress the appetite and help with weight loss after the first pregnancy.

3. Taking three to five wheat grass capsules for three to six months with a glass of water every morning can also improve ovarian function. For the past 20 years, many women reported that wheat grass helps them produce more follicles during the IVF. Even menopausal women find that wheat grass helps them produce more estrogen and progesterone by providing sufficient minerals and vitamins. Women have more

energy and better sleep during menopause when they take wheat grass.

4. Men can also drink green juice, rosemary leaf tea, and take wheat grass capsules to improve sperm morphology and motility with additional benefits: their testosterone level will increase, and they can build more muscles with more energy.

5. If, after trying the above steps for 6 to 10 months without a successful pregnancy, the fallopian tubes need to be checked with a dye injections to open the tubes. An enema may also help reopen the fallopian tubes. Many cases in which women have gotten their tubes tied can be reconnected resulting in pregnancy after using enemas for three to eight months (see below).

6. Taking Chinese herbs to improve ovarian and uterine function is very effective. In order to prevent the secondary infertility, one can start Chinese herbs right after delivery of the baby. According to Chinese Medicine Theory, women can balance their hormones much easier within the first three months of delivery. An herbalist can prescribe a formula to cleanse the uterus right after the delivery and another one to

strengthen kidney and spleen function three months before another conception. Some formulas can also prevent the post-partum depression. Chinese herbs for gynecological problems are very safe. The other benefit of taking herbs is to prevent premature ovarian failure and early menopause.

7. Acupuncture can help improve ovarian and kidney function which are critical for a second or third healthy pregnancy because the gland function becomes weaker after the first pregnancy, especially in women older than 37.

8. Enemas with herbal decoctions twice to three times a week are very effective to treat tubal blockage, fibroids, and endometriosis. It is a common, routine treatment for infertility in most Chinese Hospitals.

9. Micro-electrical current therapy with an acupuncturist: cooking Chinese herbs Tao Ren, Zao Jiao Ci, Bai Jiang Cao, and Hong Teng to make a concentrated solution; soaking gauze pieces with the liquid; then applying electrodes on the gauze attached to uterus and ovarian areas. The micro-current will help infiltrate the active ingredients of herbal decoction to the uterus and

ovarian areas. This is also a very common treatment in China.

10. Acupressure on the following points:

CV4: 4 finger widths down from belly button on the midline of abdominal area.

Uterus: one thumb width down below the CV4 on the midline of the lower abdominal area, 4 finger width lateral to the midline of the lower abdominal area.

Kid3: on the inside surface of the ankles, between the Achilles tendon and the top of the medial ankle bone

Kid5: in the tender spot, one thumb directly below Kid3.

Kid7: 3 finger widths directly above the Kid3 behind the shin bone.

11. Moxibustion is also effective on the following points:

SP8: On the inner aspect of the shin, halfway between the ankle bone and the knee (when flexed); this point is almost touching the shin bone.

SP10: When the knee is flexed and the pointer finger is touching the lower border of the kneecap, the point is under the thumb in the tender spot on the slightly inner aspect of the thigh above the knee.

SP6: palm facing down, with the pinky of the hand lying on top of the inside ankle bone, this point is located four finger widths above the ankle bone in the tender spot behind the shin.

Caution must be taken to avoid burns.

12. Use Ear Pressing Balls with herbal seeds (Wang Bu Liu Xing) on the following ear points: Kidney, Liver, Ovary, Shen Men, Uterus, Spleen.

Case:

Michelle was one of the five children in a big family. She was only 18 when she fell in love and got pregnant with her first child. Despite their young age, Michelle and her boyfriend married and started a life together adding three more children and her husband's growing business. Her husband began drinking, his alcohol abuse eventually turned into spousal abuse, so they divorced.

Eventually, she became a teacher and remarried. She was heartbroken for her new husband that she had had her tubes tied after the birth of her fourth child; the new couple wanted to have a child together. She tried two IVF

150

procedures resulting in two miscarriages, probably due to low progesterone and stress over her divorce. After a procedure to reconnect her fallopian tubes, Michelle started acupuncture and herbs to help with implantation. After only two months, she got pregnant, but it was ectopic due to the scar tissue in her fallopian tubes. Her acupuncturist told her to use herbal enema to help open the tubes and reduce the scar tissue if she was determined to have another baby.

Michelle was very determined. She cooked herbs and bought an enema kit. She started with a 100cc herbal decoction enema, and the solution stayed in her intestine over night. After a few times, she increased her herbal enema volume to 200cc. In the meantime, she continued with the oral herbs and infertility acupuncture protocol for two months.

After a few months, her tubes had opened up, and she was pregnant. The herbs and acupuncture also helped keep the pregnancy smooth and healthy. Now, she has five children, happily juggling work and a busy family. Sometimes, sudden hormone drops made her very anxious about her children and work, so her acupuncturist gave her herbs to relax and tonify her kidney and spleen, which worked well for her.

Chapter Thirteen

Post-Partum Depression

Depression is a serious mental health problem with symptoms such as low mood, fatigue, weight change, disturbed sleep, and inability to enjoy daily life despite positive aspects of life such as family or a good job. It can significantly interfere with the quality of life and even lead to suicidal thoughts. Conventional Western Medicine focuses on symptom treatment without getting to the root of the problem, and the medications prescribed also come with several side effects such as severe mood swings, insomnia, weight gain and migraines.

Why do people feel depressed?

In order to feel happy and enjoy the normal routines of life, the body needs to have chemical balance. Serotonin, GABA, and progesterone usually relax, while dopamine, glutamate, aspartate and adrenaline stimulate the brain to motivate and reward. Endorphins are also happy hormones. The stimulating, relaxing, and rewarding chemicals need to be balanced in order to achieve a good mood. If GABA, serotonin, and progesterone are too high, people feel tired and unmotivated to accomplish or achieve. The relaxation hormones help with deep sleep and digestion, so people who relax more tend to live longer; they may not be able to achieve very much, but they are happier.

On the other hand, as adrenaline and dopamine stimulate certain areas of the brain, focus and motivation are improved. Coffee and chocolate seem to stimulate the release of dopamine and adrenaline, so that women feel like they have more energy to achieve something. But those stimulating chemicals can increase heart rate and burn more calories, making the energy flow to the big muscles instead of the digestive system. Coffee also induces insulin release, so too much coffee actually can make women gain weight because high insulin moves the blood sugar into the fat cells; the body can not burn the calories but instead will store the energy as fat tissue. That is why, once people start using insulin, they tend to accumulate fat in the belly area.

High insulin can also cause inflammation and stimulate the breast cysts and prostate to grow.

When women get pregnant, they have a much higher level of progesterone; they tend to sleep more deeply and be happy with everything around them. The deep sleep and happy mood make women's immune function stronger so that they can protect their babies. Many times, women eat and drink whatever they love during pregnancy such as ice cream, cakes and spicy food. This alters their hormone balance during the pregnancy and after the delivery of the baby. Naturally after delivering a baby, progesterone levels drop to a much lower level, predisposing women toward being depressed, sad, and feeling hopeless.

In the past, women, especially in China, had whole foods and different kinds of herbs during the first three months after they delivered the baby. They did not drink much coffee, and they were told to rest as much as possible right after delivery. The post partum depression was not as bad as what women have today.

Now women need to go back to work as well as taking care of the baby at night. The lack of sleep can cause more deficiency of serotonin, GABA, and progesterone. If they do not eat sufficient fruits and vegetables, their nervous system tends to get out of balance. Also, the high stress can deplete the relaxation hormones even more, and the post-partum depression becomes more severe. The imbalance of the neurotransmitters not only can influence mood but also can change behavior and thinking. This imbalance is partly to blame for the increase of suicide and violence over the past 20 years in the general population.

Many people may still remember that a Texas mother killed her five children years ago in a very brutal way. Normal people could not understand how she could become so vicious. If she did not want her five children, why did she spend all her energy, time and money raising them? After all these years of researching about hormone balance and treating menopausal women with acupuncture and Chinese herbal medicine, I realized that hormone imbalance had caused her deep depression and hallucinations.

After giving birth to her five children, this woman did not work outside of the house or interact with other people. Her depression became so severe that she had to take medication. However, anti-depressant medications do not address the root cause of the problem: the imbalance of estrogen, progesterone, and the associated chemicals such as serotonin, dopamine, and GABA. If she had worked part-time, joined some clubs, or done volunteer work, she could have found a way to make herself feel better and to think correctly. Her mind was disturbed by this chaos of imbalanced chemicals, and she started hallucinating that if she killed all the kids, everything would get better. I strongly recommend women to take care of themselves after delivery to avoid such a tragedy.

How to prevent post partum depression?

1. Get enough sleep. Sleep can always help the body produce more relaxation hormones; depressed people do not sleep enough or do not sleep at the right time. People tend to be depressed when they go to bed after 12 am and wake up after 10 am. People who go to bed early and get up with the sunrise tend to be very positive. If the baby wakes up every two hours during the first few post partum months, allow the husband or other relatives help. Interrupted sleep can influence the nervous system, leading to anxiety and depression.

2. Put Shan Yao 30g, Gou Qi Zi 20g, Hang Qi 15g, and Dang Gui 15g in soup to improve the digestion and strengthen kidney function. Healthy digestive and kidney function can help produce happier hormones such as dopamine and progesterone.

3. Avoid spicy food, which can compromise uterine and ovarian function. Some women eat so much spicy food that they develop inflammation in their ovaries and uterus, which lowers the production of progesterone and leads to premature ovarian failure.

4. Go easy with coffee. One cup may help with energy while not interrupting sleep, but a second cup will not have the same effect. Even though the coffee does not influence sleep duration, it may change how deep the sleep is, which can cause an imbalance of the nervous system and immune function. Coffee can also move the energy away from the thyroid and adrenal glands, which can cause fatigue even with eight hours of sleep.

5. Start a part-time job or volunteer activity three months after delivering the baby. A mother's brain need to focus on something besides just taking care of the baby. Volunteering and helping people helps

increase the rewarding feelings and induces the dopamine release.

6. Avoid alcohol. Even though it can stimulate a temporary dopamine release, it does not produce more dopamine and serotonin in the body. After this huge dopamine release, depression follows the next morning. To regain the good feeling of dopamine, more alcohol is consumed, and the cycle continues until the body cannot produce dopamine at all, which is why the suicide rate is high among alcoholics. Also, deep sleep is prevented by consumption of alcohol. It causes inflammation and liver damage, and women tend to develop an autoimmune disease if they drink too much after baby's delivery. Using the temporary fix of alcohol or medications to balance hormones does not help women's depression in the long run.

7. Take a walk on the beach or in the woods where there is higher concentration of oxygen which can help balance the parasympathetic and sympathetic nerves. Taking a break from a busy life for a few days, even without traveling, will help the body produce more "happy" hormones.

8. Organic white ginger pear tea sold in Trader Joe's seems to make many women relax. It can also help increase the quality of sleep. Instead of coffee, one could drink black or green tea in the morning to get going and white tea in the afternoon to calm and relax the body for evening.

9. Take some patented Chinese herbs (the pill form) such as Jia Wei Xiao Yao Wan and Chai Hu Shu Gan San or find an herbalist to prescribe a specific formula tailored to one's specific physical and mental condition. Herbs balance the hormones and improve the function of the thyroid, adrenal glands, and endocrine and digestive systems.

10. Electrical acupuncture can help. It has been used for depression for the past 20 years. Research has verified that acupuncture can help the body produce more dopamine, serotonin, and endorphins. When women have regular acupuncture treatments, they can cut down on medications, and they can lose weight because acupuncture also can increase their metabolism and help them produce more energy. One side effect of anti-depressant medications is that they slow down the metabolism, so that people

gain weight over the years. A new study published in *Med Sci Monit.* 2017 by Ding Luo et al. indicates that electro acupuncture can modify the glutamatergic neurotransmission system to reduce the symptoms of major depressive disorders. According to this study, electro-acupuncture can be used to achieve an antidepressant effect by enhancing the function of glial glutamate transporter EAAT2, which supports 90% of glutamate neurotransmission in the hippocampus and pre-frontal cortex. However, it takes slightly longer for acupuncture to change the gene expression of the glutamate transporter EAAT2, so, when acupuncture is used to treat depression, twice to three times a week treatments are necessary to accumulate the anti-depressant effects.

11. Join some social activities. When interacting with people, the body produces more oxytocin, a "happy" hormone. Research shows that when couples get injections of oxytocin, they can think about things in a more positive way and show love to each other even though the "romantic" love has been long gone after years of marriage.

12. An optimal level of exercise can also help the body produce more endorphins. But too much intense exercise can deplete the adrenal gland and lead to depression. 15 to 30 minutes of fast walking or jogging are enough to clear up the stress hormones in the body and rewire the brain, so that there are more positive circuits than negative ones.

13. Join a singing group. Singing can instantly bring up the oxygen and nitric oxide levels as well as calm the sympathetic nervous system. Singing can also rewire the brain to clear out unhappy thoughts.

14. Having regular sex improves the production of estrogen, progesterone, and testosterone and slows down the aging of skin and ovaries. Sex can also help produce more dopamine to cope with stress. The bonus is that it is free.

15. Eating avocado a few times a week helps calm the nervous system. When one is calm, stress can be dealt with more efficiently.

16. Add apple cider vinegar to salads or put daikon or wheat bran into soups. Ginger and daikon can help the digestive system absorb vegetables and meat

more efficiently which provides better nutrients for hormone production.

17. Eat one banana in the afternoon every day to provide the body with enough tryptophan, the precursor of serotonin. When your body has enough tryptophan, carbohydrate cravings are reduced.

18. Take magnesium powder after dinner to improve sleep and relax muscles. Another option is to absorb the magnesium in a warm Epsom salt foot bath after dinner. Not only does the magnesium get absorbed through the skin with the benefits of taking magnesium powder, but it also dilates the blood vessels in the feet, drawing the blood circulation away from the head, thereby "closing up shop" on a busy mind for the night.

19. Try to avoid multi-tasking which stimulates adrenaline release and the sympathetic nervous system. Women tend to be overwhelmed when they are doing too many things at the same time. Relaxation hormones get depleted when doing more than two things at the same time. Happy women tend to do one thing at a time.

What to do if you think you have post partum depression?

1. Go see the doctor or TCM practitioner. It is important to be evaluated and rule out other conditions or problems.

2. Remember that depression is not one's fault. It is a medical condition, and no shame or guilt should be attached to having it and seeking help. Hormone imbalances, not laziness or apathy, are the root cause of depression. People seek treatment all the time for digestive or pain issues; seeking help for depression should be no different.

3. Know that there are options. Whether through TCM, conventional medicine, or lifestyle help, there are many treatment choices and combinations of choices available. Depression is not a permanent, hopeless condition. It will get better with the right help.

4. Call 911 or the Suicide Prevention Lifeline at 1-800-274-8255, open 24 hours, seven days a week if

contemplating hurting oneself or loved ones. Connecting with experts and counselors can help in a time of crisis to prevent catastrophe.

Chapter Fourteen

TCM Diagnoses for Infertility

For thousands of years, Traditional Chinese Medicine, or TCM, has been treating women for infertility and other gynecological disorders. Around 200 AD, one of the earliest TCM manuscripts is *Jin Kui Yao Lue* or *Essentials from the Golden Cabinet,* which dedicated three chapters to the discussion of gynecology. The first Obstetrics book was written in 618 AD, called *Chan Bao,* or *Deliver Treasure,* which includes the process for delivering a baby. The first pharmacology book for women's formulas followed in 936 AD; many of those formulas are included in this text and still used today. The first book of gynecology in the world was written by Dr. Qing Zhu Fu, called *Nu Ke*, or *Woman's Department* in 1636. It is no wonder that TCM, with its long history of treating gynecological disorders, has much to offer even with modern technologies available.

166

What are Some Traditional Chinese Medicine (TCM) Diagnoses and Treatments?

Chinese medicine is a system that uses patterns for diagnosis. These look at the person as a whole, not just a list of symptoms. Because this form of medicine treats the underlying cause of the body's dysfunction, it is often successful. It is difficult to diagnose oneself, so a true diagnosis must be made by a qualified Chinese medicine or TCM practitioner. Chinese herbs and food therapy are fantastic at enhancing an individual's ovarian and uterine function because they treat a person's underlying problem. The herbs should never be taken alone, but in a formula constructed by a qualified physician. The patterns themselves have names associated with organs, but do not indicate there is a physical problem with that organ; it is just a nomenclature. Some of the TCM patterns and treatments are listed for more information.

Kidney Yang deficiency:

Symptoms: delayed period, scanty menses, light red color, post-menstrual low back pain and abdominal pain relieved by a heating pad, low sex drive, dark circle around eyes, frequent night urination, loose stools, cold feet, low energy, aversion to cold, poor circulation on the lower part of the body. Allergies start in early childhood.

Treatment:
Eat warm food, avoid ice-cold drinks.
Eat ginger root, lentils, grains like oats.
Fruits should include dates and cherries.

Commonly used herbs are: Hu Lu Ba, Shu Di, Fu Zi, Rou Gui, Ai Ye, Ba Ji Tian, Du Zhong, Bu Gu Zhi, Shu Di, Sang Ji Sheng, Tu Si Zi, Xu Duan, Xian Mao, Xian Ling Pi, Zi Shi Yeng, Jin Ying Zi. Some clinical reports indicate that Fu Zi combined with Shu Di enhances ovulation.

Kidney Yin deficiency:

Symptoms: Scanty or delayed period with bright red color, insomnia, dry eyes, mouth and skin, lower back pain with

168

weak knees, poor or declining night vision, hot flashes, night sweats, low back pain with a warm sensation, heat sensations in the palms and bottoms of the feet. The tongue is bright red and dry with no coating. The pulse is thin, weak and fast.

Treatment:

Add Lycium seed and Dang Gui into soup to tonify the kidney Yin and blood.

Commonly used herbs are: Niu Zhen Zi, Han Lian Cao, Sang Ji Sheng, Shu Di, Gou Ji berry, Shan Zhu Yu, Shan Yao, Lu Jiao Jiao, Di Gu Pi.

Spleen Qi deficiency:

Symptoms: copious period, but thin and light colored, dizziness, fatigue, spotting before or after the period, abdominal pain after the period, indigestion, bloating, gas, fatigue, tired after a big meal, weak muscles and tendons, and a pale complexion. The tongue is pale or normal colored with teeth marks on the sides. The pulse is thin and weak.

Treatment:

Consume mostly organic vegetables, lightly cooked, steamed or stir fried.

Do not eat too many raw and cold foods. Don't consume ice-cold beverages, ice cream or popsicles, which can damage digestive function even more.

Cold foods should be eaten only in moderation, which include many fruits like mango, watermelon, pears, and persimmons.

Do not eat refined carbohydrates like white bread or pasta. Avoid sugar and sugar substitutes or any concentrated sweets including honey or maple syrup.

Eat grains like brown rice, Job's tears, barley, and sorghum to supplement the spleen.

Add yams, pumpkin and pumpkin seeds in soups.

Drink moderate amounts of organic whole milk which can help balance the hormones.

Commonly used herbs are: Huang Qi, Dang Shen, Bai Zhu, Fu Ling, Tai Zi Shen and Zhi Gan Cao to strengthen digestive functions.

Blood Stasis:

Symptoms: scanty, black or very dark menses with clots and stabbing pain before or during the period, many varicose veins in lower legs, fixed pain in certain body areas, some dark spots on the tongue and enlarged sublingual veins. The pulse is not smooth.

Treatment:

Add the following foods or supplements: seaweed, safflower, fish oil, OPC/pycnogenol, fresh ginger and garlic which are natural blood thinners, invigorating blood circulation.

Local application of heating pad during and after the period.

Commonly used herbs are: Tao Ren, Hong Hua, Wang Bu Ling Xing, Sang Leng, Er Zhu, Mei Gui Hua, Niu Xi, Ru Xiang, Mo Yao, Su Mu, Sheng Pu Huang, Wu Ling Zhi, Chuan Shan Jia, Zao Jiao Ci, Liu Ji Nu, Yi Mu Cao, Chong Wei Zi.

Blood Deficiency:

Symptoms: scanty, thin period with cramping or headache after it finishes, or no period, pale skin, nail beds, tongue, weakness or fatigue, palpitations, insomnia, dry hair, nails, and eyes, muscle cramps, numbness, or clicking joints or tendons, with a weak, thin or choppy pulse.

Treatment:

Add foods such as rice, wheat, red-colored vegetables, spinach, fennel, meats, red or purple fruits like dates and raisins, and dark, green, leafy vegetables.

Avoid spicy, warm or drying foods such as black tea, coffee, garlic, ginger, pepper, curry, cinnamon, and alcohol. Avoid overly cold foods such as ice cream, iced beverages, and raw vegetables.

Commonly used herbs are: Dang Gui, Shu Di Huang, Bai Shao, Ji Xue Teng, Gui Yuan, Sang Ji Sheng, He Shou Wu, Dang Shen, E Jiao, Long Yan Rou, Da Zao and Suan Zao Ren.

Liver Qi Stagnation due to high stress:

Symptoms: tendency to sigh often, chest tightness, tender sensation on the hypochondriac area, mood swings, easily angered, bitter taste in the mouth, irregular periods with an irregular temperature chart, cramping before or during the period, dark menses with breast tenderness.

Treatment:

Add foods to facilitate Liver Qi movement such as peppermint, rosemary, spearmint, and thyme.

Deep breathing exercises and meditation.

Massaging acupuncture points LV3 and LI4.

To calm the sympathetic nervous system, take magnesium powder called Calm.

Eat frequent, small meals to stabilize blood sugar levels and inhibit the release of adrenaline.

Commonly used herbs are: Chai Hu, Xiang Fu, Zhi Qiao, Chi Shao, Chuan Xiong, Qing Pi, Yan Hu Suo, Mei Gui Hua, Chuan Lian Zi, Bo He, Wu Yao, Guang Mu Xiang.

Phlegm condition:

Symptoms: tends to develop cysts in the ovaries or other parts of the body, clears throat frequently, bronchitis, asthma or sinusitis several times during the winter, may have lipoma in different parts of the body, foggy headed, tends to sleep more but feel tired in the morning; feels better after exercise.

Treatment:

To reduce phlegm, consume fresh ginger.

Avoid dairy products and greasy foods, as well as salty or sweet foods.

Commonly used herbs are: Dang Shen, Bai Zhu, Cang Zhu, Fu Ling, Chen Pi, Bai Zhi, Fa Ban Xia, Dan Nan Xing, Zhe Bei Mu.

Damp Heat:

Symptoms: tends to have a lot of thick, yellowish vaginal discharges, lower leg swelling, frequent dark urination, easily develops urinary tract infections, a heavy sensation in the lower legs, ovarian cysts, smelly and sticky stools.

Treatment:

Avoid spicy, salty, sweet or greasy foods.

Commonly used herbs are: Niu Xi, Bai Zhi, Pu Gong Ying, Huang Bai, Huang Qin, Sheng Di Yu, Cang Zhu, Ma Bian Cao, Wan Can Sha, Chun Gen Bai Pi, Lu Xuan Cao, Bai Jiang Cao, Yu Xing Cao, and Cao He Che.

In Summary:

Most herbs used in treating women's problems are very safe with few side effects if used properly. By balancing the hormones, one not only enhances fertility but also menopause can be postponed or menopausal symptoms reduced or eliminated with herbs.

There are other diagnoses in TCM which are not included in this list. One diagnosis does not usually cover all the patient's symptoms, and a qualified TCM practitioner can tailor the diagnosis to the patient to maximize results. Along with diet and herbs, regular treatments with acupuncture can help treat the underlying cause of fertility issues.

References

Barad, D.H. and Gleicher, N. Increased Oocyte Production After Treatment with Dehydroepiandrosterone. *Fertil. Steril.* 2005. 84:756.

Chen, B.Y. and Yu, J. Relationship Between Blood Radioimmunoreactive Beta-endorphin and Hand Skin Temperature During the Electro-acupuncture Induction of Ovulation. *Acupunct. Electrother. Res.* 1991. 16:1-5.

Chen, D., Chen, S.R., Shi, X.L., et al. Clinical Study on Needle-pricking Therapy for Treatment of Polycystic Ovarian Syndrome. *Zhongguo Zhen Jiu.* 2007. 27:99-102.

Daya, S. 20-25% of miscarriages are due to immune system problems. *Chatelaine Magazine.* November, 1993. www.chem-tox.com/infertility/

Dean, C.M. *Hormone balance: A Woman's Guide to Restoring Health and Vitality.* 2005. Adams Media Corporation.

Feodor Nilsson, S., Andersen, P.K., Strandberg-Larsen, K., et al. Risk factors for Miscarriage from a Prevention Perspective: A Nationwide Follow-up Study. *BJOG.* 2014. 121:1375-1384.

Gong, D., Liang, C., Lai, X., et al. Effects of Different Acupuncture Manipulation on Plasma Estradiol, Testosterone and Cortisol in Patients with Kidney Deficiency. *Zhen Ci Yan Jiu.* 1993.18:253-256.

177

Hotze, S. and Griffin, K. *Hormones, Health, and Happiness, A Natural Medical Formula for Rediscovering Youth*. 2005. Forrest Publishing.

Khan, K.N., Masuzaki, H., Fujishita, A., et al. Estrogen and Progesterone Receptor Expression in Macrophages and Regulation of Hepatocyte Growth Factor by Ovarian Steroids in Women with Endometriosis. *Hum. Reprod.* 2005. 20:2004-2013.

Kristensen, D.M., Desdoits-Lethimonier, C., Mackey, A.L., et al. Ibuprofen Alters Human Testicular Physiology to Produce a State of Compensated Hypogonadism. *Proc. Natl. Acad. Sci. USA*. 2018. Jan 8. pii: 201715035. doi:10.1073/pnas.1715035115.

Liu, X.Y., Huang, G.Y., and Zhang, M.M. Preliminary study on the mechanisms of acupuncture in promoting embryo implantation in rats. *Zhongguo Zhong Xi Yi Jie He Za Zhi*. 2007. 27:633-636.

Luo, D., Ma, R., Wu, Y., Zhang, X, Liu, Y., Wang, L., and Fu. W. Mechanism Underlying Acupuncture-Ameliorated Depressive Behaviors by Enhancing Glial Glutamate Transporter in Chronic Unpredictable Mild Stress (CUMS) Rats. *Med Sci Monit. 2017*. 23:3080-3087.

Negro, R., Formoso, G., Coppola, L., et al. Euthyroid Women with Autoimmune Disease Undergoing Assisted Reproduction Technologies: The Role of Autoimmunity and Thyroid Function. *J. Endocrinol. Invest*. 2007. 30:3-8.

Paulus, W.E., Zhang, M., Strehler, E., et al. Influence of Acupuncture on the Pregnancy Rate in Patients Who Undergo Assisted Reproduction Therapy. *Fertil. Steril.* 2002. 77:721-724.

Siterman, S., Eltes, F., Wolfson, V., et al. Does Acupuncture Treatment Affect Sperm Density in Males with Very Low Sperm Count? A pilot study. *Andrologia.* 2000. 32:31-39.

Stener-Victorin, E., Lundeberg, T., Waldenström, U., et al. Effects of Electro-acupuncture on Nerve Growth Factor and Ovarian Morphology in Rats with Experimentally Induced Polycystic Ovaries. *Biol. Reprod.* 2000. 63:1497-1503.

Stener-Victorin, E., Waldenström, U., Tägnfors, U., et al. Effects of Electro-acupuncture on Anovulation in Women with Polycystic Ovary Syndrome. *Acta. Obstet. Gynecol. Scand.* 2000. 79:180-188.

Thrasher, J.D., Madison, R., and Broughton, A. Immunologic Abnormalities in Humans Exposed to Chlorpyrifos: Preliminary Observations. *Arch. Environ. Health.* 1993. 48:89-93.
Wolfgang in Germany:

Index

www.ingramcontent.com/pod-product-compliance
Lightning Source LLC
Chambersburg PA
CBHW032006170526
45157CB00002B/573